Dairy Science and Technology

Dairy Science and Technology

Gavin White

RCALLISTO
REFERENCE

www.callistoreference.com

Callisto Reference,
118-35 Queens Blvd., Suite 400,
Forest Hills, NY 11375, USA

Visit us on the World Wide Web at:
www.callistoreference.com

ISBN: 978-1-64116-534-1 (Hardback)

Cataloging-in-Publication Data

Dairy science and technology / Gavin White.
     p. cm.
Includes bibliographical references and index.
ISBN 978-1-64116-534-1
1. Dairying. 2. Milk. 3. Dairy farming. 4. Dairy processing--Technological innovations.
5. Dairy products. I. White, Gavin.
SF239 .D35 2022
636 214 2--dc23

TABLE OF CONTENTS

This book aims to help a broader range of students by exploring a wide variety of significant topics related to this discipline. It will help students in achieving a higher level of understanding of the subject and excel in their respective fields. This book would not have been possible without the unwavered support of my senior professors who took out the time to provide me feedback and help me with the process. I would also like to thank my family for their patience and support.

Dairy refers to the business enterprise which deals with the harvesting and processing of animal milk from animals such as cows, goats, buffaloes and sheep. The branch of science which focuses on the production and manufacturing of various dairy products such as milk, cheese, yogurt and butter is termed as dairy science. It is also involved in the study of animal nutrition, animal reproduction and lactation. Dairy technology is a branch of engineering which explores the processing of milk and milk products. It also deals with the packaging, transportation and distribution of dairy products through applying principles from the fields of nutrition, bacteriology and biochemistry. This book provides significant information of this discipline to help develop a good understanding of dairy science and technology. While understanding the long-term perspectives of the topics, it makes an effort in highlighting their impact as a modern tool for the growth of the discipline. Those in search of information to further their knowledge will be greatly assisted by this book.

A brief overview of the book contents is provided below:

Chapter – Introduction

The technology and science which focuses on the production of milk and milk products, such as cheese, yogurt, butter, ice cream and casein, is referred to as dairy science and technology. This is an introductory chapter which will introduce briefly all the significant aspects of dairy science and technology.

Chapter – Milk: Types and Properties

The liquid which is secreted by the mammary glands of female mammals in order to nourish their young ones for period beginning immediately after birth is known as milk. There are various types of milk such as raw milk, organic milk, baked milk, malted milk, condensed milk, scalded milk and soy milk. This chapter has been carefully written to provide an easy understanding of these types of milk.

Chapter – Dairy Products

Some of the commonly used dairy products include yogurt, powdered milk, cream, butter, custard, ice cream, cheese, fermented milk, clabber, soured milk, kefir, kumis, viili and buttermilk. The topics elaborated in this chapter will help in gaining a better perspective about these dairy products.

Chapter – Dairy Processing

Dairy processing includes various processes and methods such as thermization, milk powder production, clarification and cream separation, sterilization, homogenization, fluid milk production, pasteurization, etc. This chapter closely examines these techniques of dairy processing to provide an extensive understanding of the subject.

Chapter – Technologies used in Dairy Industry

Various technologies are used in dairy industry such as separators, homogenizers and packing machines for milk. There are also various cheese making, butter making and ice-cream making equipment that are used in the dairy industry. This chapter discusses in detail these technologies and equipment related to dairy industry.

Gavin White

Introduction

The technology and science which focuses on the production of milk and milk products, such as cheese, yogurt, butter, ice cream and casein, is referred to as dairy science and technology. This is an introductory chapter which will introduce briefly all the significant aspects of dairy science and technology.

DAIRY

Milk is a bundle of nutrients, all contained in a nondescript white liquid. Although milk's presence as a beverage at meals may not be as popular as it used to be, milk is used in many products that are consumed throughout the day.

On the Food Guide Pyramid, milk and dairy products are placed near the top because, although they are part of a healthful diet, they should be consumed in moderation. Adults should consume 2 servings of low-fat or nonfat dairy products daily; 1 serving equals 1 cup of milk or yogurt or 1 1/2 ounces of cheese. Children and pregnant or lactating women should add an extra serving each day. Milk and other dairy foods are rich in calcium, a mineral important for developing strong bones and teeth and for nerve transmission. They are also an important source of many vitamins and minerals. Large quantities of these foods, however, are not needed to ensure that you are getting adequate amounts of these nutrients. Just three 8-ounce glasses of skim milk, for example, provide nearly all of the calcium you need each day.

Some people do not include enough dairy foods in their diets. One reason is the mistaken belief that all dairy products are high in fat. Some are, but there is an abundance of low-fat and nonfat dairy products, from milk to yogurt to cheese.

Other people do not consume dairy foods because of intolerance to milk sugar or allergy to milk proteins. However, those with intolerance to milk often do not need to follow a diet that is completely milk-free. People with allergy to milk must avoid dairy foods and may want to get help with adjusting their diets to ensure nutritional adequacy.

Another reason people do not consume dairy products is the growing consumption of soda pop. The average American drinks about a half gallon of milk a week but, in comparison, about 11 cans or a gallon of soda pop a week. Taking calcium supplements or eating calcium-enriched food can help you obtain needed calcium, but dairy foods are an easy way to get the calcium and other essential nutrients you need.

Milk can be consumed in its fluid form, in a more solid form (such as yogurt), as cheese, or as a major ingredient that is added to other foods. Dairy cases now abound with milk-based products and their reduced-fat and nonfat versions. The cornucopia of dairy products includes the following:

- Fluid milk — Although cow's milk is generally consumed in the United States, other cultures use milk from goats, camels, llamas, reindeer, sheep, and water buffalo. Milk is a staple in diets worldwide.

- Dried and concentrated milk — These products include powdered milk, evaporated milk, and condensed milk.

- Cheese — Cheese is made by coagulating and draining milk or cream or a combination of both.

- Yogurt — Yogurt is made by adding bacteria to milk to ferment it.

- Ice cream and other dairy desserts — Ice cream and other frozen desserts are simply milk or cream to which sugar, flavorings, and, often, eggs have been added.

- Cream and sour cream — Cream is the fat that rises to the top of the surface in unprocessed milk. Sour cream is simply cream that has been fermented or thickened. The cream is usually "soured" by adding bacteria to it, much in the way that yogurt is created.

- Butter — This yellowish substance is essentially fat that has been separated from cream.

Processing of Milk and Milk Products

Virtually all milk and milk used in dairy products is pasteurized. Pasteurization is a process invented by French chemist Louis Pasteur. It uses heat to destroy harmful bacteria in milk, but it retains the nutritional value of milk.

Pasteurization kills bacteria that have been responsible for major plagues such as tuberculosis, polio, scarlet fever, and typhoid fever. It is also advantageous because it destroys many of the bacteria that cause spoilage and many of the enzymes that promote rancidity. Pasteurization, therefore, increases both shelf life and safety of milk.

A common term that consumers see when purchasing milk is "homogenized." Homogenization is a process introduced in the 1950s in which fat globules in the milk are broken down so they are evenly dispersed throughout the milk. Most milk at the supermarket is homogenized.

During homogenization, milk is forced through a small opening at high pressure. The product has a smoother, richer texture and a whiter color than nonhomogenized milk.

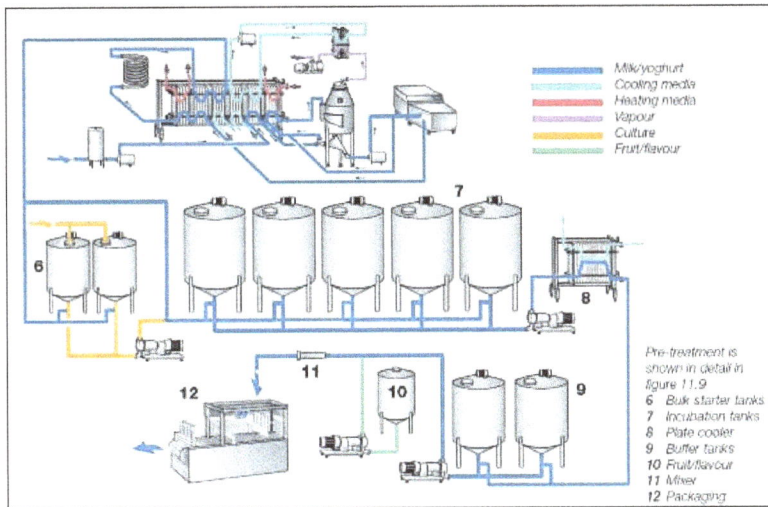

Nutrition

Milk and dairy products provide many of the key nutrients needed daily, particularly calcium.

Milk and dairy products also supply high-quality protein. Because of its animal source, milk protein is complete meaning it provides a sufficient amount of the nine essential amino acids.

Dairy products are also naturally rich in B vitamins and most of the minerals considered to be essential in the diet, including calcium, magnesium, phosphorus, zinc, iodine, and selenium. In addition, milk also contains several vitamins and minerals that have been added to meet the requirements of the Food and Drug Administration. Low-fat and nonfat milk may be fortified with vitamin A because this fat-soluble vitamin is lost when the milk fat is removed. Vitamin D is added to all milk to help the body better use calcium.

Milk also is a good source of carbohydrates. With the exception of cheeses and butter, milk products are higher in carbohydrates than protein or fat. Milk's carbohydrate is lactose, a sugar unique to milk that is actually two sugars (glucose and galactose) linked together. Food scientists call this type of sugar a "disaccharide."

Lactose is not as sweet as other sugars. It helps the body absorb calcium and phosphorus and may even help in the growth of friendly bacteria needed in the intestines. In addition, galactose, one of the sugars in lactose, is a vital part of brain and nerve tissue. It is released when the body digests lactose. Lactose is a bit of a paradox, however. Although it has these beneficial properties, many people have difficulty digesting milk.

Lactose Intolerance

As many as 50 million Americans are estimated to have lactose intolerance an inability to adequately digest ordinary amounts of dairy products such as milk and ice cream.

Worldwide, nearly 70 percent of the adult population is thought to be lactose intolerant, and the condition is very common among American Indians and those of Asian, African, Hispanic, and Mediterranean descent.

Lactose is the sugar that is naturally present in milk and milk products. It must be broken down by lactase (an enzyme found in the intestine) before the body can use it. If there is not enough lactase, undigested milk sugar remains in the intestine. Bacteria in the colon then ferment this sugar. Gas, cramping, and diarrhea can follow.

Most of us begin to lose intestinal lactase as we age. However, this occurs to varying degrees. Thus, people with lactase deficiency vary in their ability to comfortably digest milk and milk products.

As obvious as the symptoms of lactose intolerance may be, it is not easily diagnosed from the symptoms alone. Many other conditions, including stomach flu and irritable bowel syndrome, can cause similar symptoms.

Persons with milk allergies should avoid milk, but those with lactose intolerance often do not need to follow a diet that is completely lactose-free. The following suggestions may help:

- Avoid eating or drinking large servings of dairy products at one time. (Several smaller servings over the course of a few hours are much easier to digest).

- Drink milk or eat dairy products with a meal.

- Choose hard or aged cheeses, such as swiss or cheddar, over fresh varieties. Hard cheeses have smaller amounts of lactose and are more likely to be tolerated.

- Take lactase tablets or drops, such as Lactaid or Dairy Ease. These types of products contain the enzyme that breaks down lactose, reducing the amount that your body must digest on its own.

Despite all the nutrients in milk, the nutritional advantages of dairy products must be weighed against the potential health drawbacks of two key components in milk: sodium and fat. Whole milk, cream, and cheeses contain substantial amounts of fat, especially saturated fat. These fats add calories and have been tied to higher cholesterol levels and cardiovascular disease. However, it is important to note that low-fat and nonfat milk varieties are available and are significantly lower in fat than whole milk. In addition, depending on how much is consumed, milk or products made from milk may be a major source of sodium a special concern for anyone following a low-sodium diet.

Selection

Unless dried or canned, milk and dairy products are perishable. For that reason, most have an expiration date printed on the packaging. The date often states, "Sell by" and is a good indicator of freshness. Look for the date before buying and before consuming a product. Usually, dairy products will keep about a week beyond that date.

Storage

Keep milk in the coldest part of your refrigerator. Avoid storing milk in the refrigerator door unless it has a special compartment designed to keep the milk colder than in the rest of the refrigerator.

Keep yogurt and fresh cheeses in airtight containers in the refrigerator. Loosely wrapped, these foods will pick up smells in the refrigerator, possibly leaving them with an undesirable taste.

Cheeses such as cottage cheese, ricotta, and cream cheese will keep for 1 week after the sell-by date. Soft cheeses such as Brie, Camembert, Muenster, and mozzarella and blue-vein cheeses can keep from 1 to 3 weeks. Semi-firm and hard cheeses, such as cheddar and Monterey Jack, will keep as long or longer. Generally, the harder the cheese, the longer it will remain fresh when carefully stored.

Shredded cheese will not keep as long because it has more surface exposed to the air. Soft cheese that has mold on it should be discarded. Firm cheese that has mold can sometimes be used as long as 1/2 inch to 1 inch of cheese near the molded spot has been cut away and discarded. If any milk or milk product has a strange odor, throw it out.

Safety Issues

Some small markets or independent farmers still sell raw milk. Because it has not been pasteurized, this milk may contain germs that make you ill. For that reason, the sale of raw milk is often prohibited by law, depending on location.

For some people, proteins in cow's milk may trigger allergic reactions. Whey proteins (beta-lactoglobulin and beta-lact-albumin) and casein are the primary proteins that trigger allergic reactions. Symptoms of a milk allergy may include nasal congestion, hives, itching, swelling, wheezing, shortness of breath, nausea, upset stomach, cramps, heartburn, gas or diarrhea, light-headedness, and fainting.

It is easy to confuse a milk allergy with another common health concern related to dairy foods — lactose intolerance. Lactose intolerance also can lead to nausea, vomiting, cramping, and diarrhea. However, if you have lactose intolerance, you usually can eat small amounts of dairy food without problems. In contrast, a tiny amount of a food to which you are allergic can trigger a reaction.

DAIRY SCIENCE AND TECHNOLOGY

Dairy science explores the technology and science behind the production of milk and milk products like cheese, yogurt, butter, ice cream and casein. It focuses on the biological, chemical, physical, and microbiological aspects of milk itself, and on the technological (processing) aspects of the transformation of milk into its various consumer products, including beverages, fermented products, concentrated and dried products, butter and ice cream.

Milk is as ancient as humankind itself, as it is the substance created to feed the mammalian infant. All species of mammals, from humans to whales, produce milk for this purpose. Many centuries ago, perhaps as early as 6000-8000 BC, ancient peoples learned to domesticate species of animals for the provision of milk to be consumed by them. These included cows (genus Bos), buffaloes, sheep, goats, and camels, all of which are still used in various parts of the world for the production of milk for human consumption.

Fermented products such as cheeses were discovered by accident, but their history has also been documented for many centuries, as has the production of concentrated milks, butter, and even ice cream.

Technological advances have only come about very recently in the history of milk consumption, and our generations will be the ones credited for having turned milk processing from an art to a science. The availability and distribution of milk and milk products today in the modern world is a blend of the centuries old knowledge of traditional milk products with the application of modern science and technology.

The role of milk in the traditional diet has varied greatly in different regions of the world. The tropical countries have not been traditional milk consumers, whereas the more northern regions of the world, Europe (especially Scandinavia) and North America, have traditionally consumed far more milk and milk products in their diet. In tropical countries where high temperatures and lack of refrigeration has led to the inability to produce and store fresh milk, milk has traditionally been preserved through means other than refrigeration, including immediate consumption of warm milk after milking, by boiling milk, or by conversion into more stable products such as fermented milks.

Milk: Types and Properties

The liquid which is secreted by the mammary glands of female mammals in order to nourish their young ones for period beginning immediately after birth is known as milk. There are various types of milk such as raw milk, organic milk, baked milk, malted milk, condensed milk, scalded milk and soy milk. This chapter has been carefully written to provide an easy understanding of these types of milk.

MILK

Milk is the liquid secreted by the mammary glands of female mammals to nourish their young for a period beginning immediately after birth. The milk of domesticated animals is also an important food source for humans, either as a fresh fluid or processed into a number of dairy products such as butter and cheese.

Almost all the milk now consumed in Western countries is from the cow, and milk and milk products have become important articles of commerce. Other important sources of milk are the sheep and goat, which are especially important in southern Europe and the Mediterranean area; the water buffalo, which is widely domesticated in Asia; and the camel, which is important in the Middle East and North Africa.

Milk is essentially an emulsion of fat and protein in water, along with dissolved sugar (carbohydrate), minerals, and vitamins. These constituents are present in the milk of all mammals, though their proportions differ from one species to another and within species. The milk of each species seems to be a complete food for its own young for a considerable time after birth. In the stomachs of the young, milk is converted to a soft curd that encloses globules of fat, enabling digestion to proceed smoothly without the disturbance often caused by fatty food. Lactose, or milk sugar, is broken down into simpler digestible sugars by the enzyme lactase, which is produced in the intestine of infants. Infants who do not produce lactase develop lactose intolerance, a condition in which a variety of gastrointestinal problems arise. Lactose intolerance also commonly develops after weaning or with advancing age, when many individuals cease producing lactase.

Nutrient composition of the whole milk of humans and select domesticated animals (per 100 g)						
Source	Energy (kcal)	Fat (g)	Cholesterol (mg)	Protein (g)	Calcium (mg)	Phosphorus (mg)	Carbohydrate (g)
Human	70	4.38	14	1.03	32	14	6.89
Cow	61	3.34	14	3.29	119	93	4.66

Goat	69	4.14	11	3.56	134	111	4.45
Sheep	108	7.00	—	5.98	193	158	5.36
Water buffalo	97	6.89	19	3.75	169	117	5.18

Milk protein is of high nutritional value because it contains all the essential amino acids—i.e., those which infants cannot synthesize in the necessary quantities. Milk's mineral content includes calcium and phosphorus in quantities sufficient for normal skeletal development, but little iron. Milk contains B vitamins as well as small amounts of vitamins C and D. Commercial cow's milk is commonly enriched with vitamins D and A before sale.

Microorganisms contained in raw (unheated) milk or picked up from the environment will quickly sour and curdle the milk. Cooling to slightly above its freezing point keeps milk palatable for a longer time by reducing the multiplication of spoilage bacteria and the chemical changes that they induce.

Many countries have laws requiring that milk be pasteurized as a protection against pathogenic (disease-causing) organisms. Pasteurization is a partial sterilization accomplished by raising the milk to a temperature high enough to destroy pathogenic bacteria and a large proportion of those causing spoilage. Pasteurized milk that is kept refrigerated in closed containers will remain consumable for approximately 14 days.

Milk fat, being less dense than other milk components, can be efficiently removed in a cream separator by centrifugation, yielding low-fat milk and skim milk. Low-fat milk contains 1–2 percent fat, while skim milk contains less than 0.5 percent fat.

Much of the milk sold as a beverage has undergone homogenization, a process in which the milk is forced under high pressure through small openings to distribute the fat evenly throughout the milk.

RAW MILK

Raw milk or unpasteurized milk is milk that has not been pasteurized, a process of heating liquid foods to decontaminate them for safe drinking. Proponents of raw milk have stated that there are benefits to its consumption, including better flavor, better nutrition, and the building of a healthy immune system. However, the medical community has warned of the dangers, which include a risk of infection, and has not found any clear benefit. The availability and regulation of raw milk vary around the world. In the US, some dairies have adopted low-temperature vat pasteurization, which they say produces a product similar to raw milk.

Humans first learned to regularly consume the milk of other mammals following the domestication of animals during the Neolithic Revolution or the development of agriculture. This development occurred independently in several places around the world from as early as 9000–7000 BC in Mesopotamia to 3500–3000 BC in the Americas. The most important dairy animals—cattle, sheep and goats—were first domesticated in Mesopotamia, although domestic cattle had been independently derived from wild aurochs populations several times since. From there dairy animals spread to Europe (beginning around 7000 BC but not reaching Britain and Scandinavia until after 4000 BC), and South Asia (7000–5500 BC).

Pasteurization is widely used to prevent infected milk from entering the food supply. The pasteurization process was developed in 1864 by French scientist Louis Pasteur, who discovered that heating beer and wine was enough to kill most of the bacteria that caused spoilage, preventing these beverages from turning sour. The process achieves this by eliminating pathogenic microbes and lowering microbial numbers to prolong the quality of the beverage.

After sufficient scientific study led to the development of germ theory, pasteurization was introduced in the United States in the 1890s. This move successfully controlled the spread of highly contagious bacterial diseases including *E. coli*, bovine tuberculosis and brucellosis (all thought to be easily transmitted to humans through the drinking of raw milk). In the early days after the scientific discovery of bacteria, there was no product testing to determine whether a farmer's milk was safe or infected, so all milk was treated as potentially contagious. After the first tests were developed, some farmers took steps to prevent their infected animals from being killed and removed from food production, sometimes even falsifying test results to make their animals appear free of infection. Recent advances in the analysis of milk-borne diseases have enabled scientists to track the DNA of the infectious bacteria to the cows on the farms that supplied the raw milk.

The recognition of many potentially deadly pathogens, such as E. coli 0157 H7, Campylobacter, Listeria, and Salmonella, and their possible presence in poorly produced milk products has led to the continuation of pasteurization. The Department of Health and Human Services, Center for Disease Control and Prevention, and other health agencies of the United States strongly recommend that the public do not consume raw milk or raw milk products. Young children, the elderly, people with weakened immune systems, and pregnant women are more susceptible to infections originating in raw milk.

Milk can be repasteurized, as is done when pasteurized milk is shipped from the US mainland to Hawaii, which can be done to extend the expiration date.

Health Effects

Infectivity

The potential pathogenic bacteria from raw milk, include tuberculosis, diphtheria, typhoid, and streptococcal infections, make it potentially unsafe to consume. Raw milk poses a realistic health threat due to a possible contamination with human pathogens. It is therefore strongly recommended that milk should be heated before consumption.

Even with precautions and cold storage (optimally 3 to 4 °C), raw milk has a shelf life of 3 to 5 days.

Nutrition and Allergy

With the exception of an altered organoleptic flavor profile, heating (particularly ultra high temperature and similar treatments) will not substantially change the nutritional value of raw milk or other benefits associated with raw milk consumption.

Raw milk advocates, such as the Weston A. Price Foundation, say that raw milk can be produced hygienically, and that it has health benefits that are destroyed in the pasteurization process. Research shows only very slight differences in the nutritional values of pasteurized and unpasteurized milk.

Three studies have found a statistically significant inverse relationship between consumption of raw milk and asthma and allergies. However, all of these studies have been performed in children living on farms and living a farming lifestyle, rather than comparing urban children living typical urban lifestyles and with typical urban exposures on the basis of consumption or nonconsumption of raw milk. Aspects of the overall urban vs. farming environment lifestyle have been suggested as having a role in these differences, and for this reason, the overall phenomenon has been named the "farm effect". A recent scientific review concluded that "most studies alluding to a possible protective effect of raw milk consumption do not contain any objective confirmation of the raw milk's status or a direct comparison with heat-treated milk. Moreover, it seems that the observed increased resistance seems to be rather related to the exposure to a farm environment or to animals than to raw milk consumption." For example, in the largest of these studies, exposure to cows and straw as well as raw milk were associated with lower rates of asthma, and exposure to animal feed storage rooms and manure with lower rates of atopic dermatitis; "the effect on hay fever and atopic sensitization could not be completely explained by the questionnaire items themselves or their diversity".

Epidemiology

Before purified milk was adopted in the US, public health officials were concerned with cow milk transmission of bovine tuberculosis to humans with an estimated 10% of all tuberculosis cases in humans being attributed to milk consumption. Along with specific diseases, officials continue to be concerned about outbreaks. With the use of modern pasteurization and sanitation practices milk accounts for less than 1% of reported outbreaks from food and water consumption. As comparison, raw milk was associated with 25% of all disease outbreaks from food/water during the time before World War II in the U.S. From a public health stand point pasteurization has decreased the percentage of milk associated food/water borne outbreaks.

Outbreaks have occurred in the past from consuming food products made with raw milk. One of the potential pathogens in raw milk, Listeria monocytogenes, can survive the pasteurization process and contaminate post-pasteurization environments. Milk and dairy products made with that milk then become recontaminated. Consistent contamination persists by bacteria survival in biofilms within the processing systems. One food item that has commonly used raw milk in its production in the past is cheese. Several different types of cheeses made with raw milk are consumed by a large portion of the United States population, including soft cheeses. Since Gouda cheese has a 60-day aging period prior to its consumption, it has previously been hypothesized that no bacteria would persist through that time.

Between 2007 and 2016 there were 144 outbreaks connected to raw milk consumption in the United States. Because raw milk production skips the pasteurization process, the germs that are normally removed remain in the milk product. Exposure to raw milk containing harmful germs poses a threat of infection, resulting from bacteria including Camplyobacter, Cryptosporidium, E. coli, Listeria, and Salmonella. Additionally, depending on the severity of infection, there may be further threat to human health. Infection has the potential to induce serious illness such as Guillain-Barré syndrome and Hemolytic Uremic Syndrome (HUS). Because of the vulnerability of developing and degrading immune systems, children, pregnant women, the elderly, and those who are immunocompromised are at a heightened risk of experiencing infection from raw milk consumption.

One study used mice to evaluate the difference in nutritional values between raw and pasteurized milk. Mice were separated into two groups: a pasteurized milk group and a raw milk group. Each group consisted of breeding pairs. The conclusion of the study measured no significant difference in weights of pasteurized to raw milk consuming mice. Birth weights were measured from each group and showed no significant differences between groups. Overall the study showed no measurable significant difference in nutritional value in growth and fertility of mice.

Uses

Raw yak milk is allowed to ferment overnight to become yak butter. Some cheeses are produced with raw milk although local statutes vary regarding what if any health precautions must be followed such as aging cheese for a certain amount of time.

A thick mixture known as Syllabub was created by milkmaids squirting milk directly from a cow into a container of cider, beer, or other beverage.

ORGANIC MILK

A glass of milk.

Organic milk refers to a number of milk products from livestock raised according to organic farming methods. In most jurisdictions, use of the term "organic" or equivalents like "bio" or "eco", on any product is regulated by food authorities. In general these regulations stipulate that livestock must be: allowed to graze, be fed an organically certified fodder or compound feed, not be treated with most drugs (including growth hormone), and in general must be treated humanely.

There are multiple obstacles to forming firm conclusions regarding possible safety or health benefits from consuming organic milk or conventional milk, including the lack of long term clinical studies. The studies that are available have come to conflicting conclusions with regard to absolute differences in nutrient content between organic and conventionally produced milk, such as protein or fatty acid content. The weight of available evidence does not support the position that there are any clinically relevant differences between organic and conventionally produced milk, in terms of nutrition or safety.

Comparison with Conventional Milk

Chemical Composition

Studies have examined chemical differences in the composition of organic milk compared with conventional milk. These studies generally suffer from confounding variables, and are difficult to generalize due to differences in the tests that were done, the season of testing and brand of milk tested, and because the vagaries of agriculture affect the chemical composition of milk. Treatment of the foodstuffs after initial gathering (whether milk is pasteurized or raw), the length of time between milking and analysis, as well as conditions of transport and storage, also affect the chemical composition of a given batch.

Nutrient Content

A 2012 meta-analysis of the scientific literature did not find significant differences in the vitamin content of organic and conventional plant or animal products, and found that results varied from study to study. The authors found 4 studies on each of beta-carotene and alpha-tocopherol levels in milk; differences were heterogeneous and not significant. The authors found few studies on fatty acids in milk; all (but for one) were of raw milk, and suggest that raw organic milk may contain significantly more beneficial omega-3 fatty acids and vaccenic acid than raw conventional milk. The authors found no significant differences between organic raw milk and conventional milk with respect to total protein, total fat, or 7 other vitamins and fatty acids tested. A different review concluded, "Results to date suggest that the nutritional content of organic milk is similar to that of conventional milk. There may be a different profile of fatty acids in organic milk, with a higher proportion of PUFA (polyunsaturated fatty acids) relative to other fatty acids, but this effect does not appear to be consistent. This difference will be smaller in fat-reduced milk".

A less comprehensive review published in 2012 looking only at data from studies published from 2008 to 2011 found that organic dairy products contain significantly higher protein, total omega-3 fatty acid, and 5 other fatty acids, but less linoleic acid, oleic acid, and omega-6 fatty acids than those of conventional produced milk. It also found that organic dairy products have significantly higher omega-3 to -6 ratio and Δ9-desaturase index than the conventional types.

Chemical and Pesticide Residue

A consumer concern that drives demand for organic food is the concern that conventional foods may contain residues of pesticides and chemicals. Many investigations of organic milk have not measured pesticide residues. One review of the literature concluded the "available evidence indicates that regular and organic milk contain similar trace levels of chemical and pesticide residues".

Health and Safety

With respect to scientific knowledge of health and safety benefits from a diet of organic food, several factors limit our ability to say that there is any health benefit, or detriment, from such a diet. The 2012 meta-analysis noted that "there have been no long-term studies of health outcomes of populations consuming predominantly organic versus conventionally produced food controlling for socioeconomic factors; such studies would be expensive to conduct." A 2009 meta-analysis has noted that there have been very few studies that have looked at direct human health outcomes. In addition, difficulties in accurately and meaningfully measuring chemical differences between organic and conventional milk make it difficult to extrapolate health recommendations based solely on chemical analysis.

The authors of the 2012 meta-analysis ultimately concluded that the review "identified limited evidence for the superiority of organic foods. The evidence does not suggest marked health benefits from consuming organic versus conventional foods".

"There is no evidence of clinically relevant differences in organic and conventional milk. There are few, if any, nutritional differences between organic and conventional milk. There is no evidence that any differences that may exist are clinically relevant. There is no evidence that organic milk has clinically significant higher bacterial contamination levels than does conventional milk. There is no evidence that conventional milk contains significantly increased amounts of bovine growth hormone. Any bovine GH that might remain in conventional milk is not biologically active in humans because of structural differences and susceptibility to digestion in the stomach".

Taste

One review noted that some consumers like the taste of organic milk, while others do not, and suggested that the amount of heat treatment is likely to be a significant factor in determining the taste of the milk. Certain treatments, such as ultra-heat treatments used by milk producers, can impart a slight nutty taste to the milk. Overall, the results of taste testing "are not clear-cut" as to whether organic or conventional milk is preferred.

Economic Factors

Compared to conventional milk farms, organic milk farms produce significantly less milk per cow and cost more to operate. Organic dairy co-ops have been a successful economic survival strategy for small to medium-sized producers in the American midwest. Organic milk accounts for 18% of milk sales in the US and was worth $2.5 billion in 2016.

BAKED MILK

Baked milk is a variety of boiled milk that has been particularly popular in Russia, Ukraine and Belarus. It is made by simmering milk on low heat for eight hours or longer.

In rural areas baked milk has been produced by leaving a jug of boiled milk in an oven for a day or for a night until it is coated with a brown crust. Prolonged exposure to heat causes reactions between the milk's amino acids and sugars, resulting in the formation of melanoidin compounds that give it a creamy color and caramel flavor. A great deal of moisture evaporates, resulting in a change of consistency. The stove in a traditional Russian loghouse (izba) sustains "varying cooking temperatures based on the placement of the food inside the oven".

Baked milk.

Today, baked milk is produced on an industrial scale. Like scalded milk, it is free of bacteria and enzymes and can be stored safely at room temperature for up to forty hours. Home-made baked milk is used for preparing a range of cakes, pies, and cookies.

Fermented Baked Milk

Ryazhenka and varenets are fermented baked milk products, a sort of traditional yoghurt. It is a common breakfast drink in Ukraine, Belarus, and Russia.

In peasant communities, the varenets has been made in the traditional East Slavic oven by "baking sour milk to a golden brown color". In the Soviet era, the name "ryazhenka" became to be applied to the government-produced creme-colored drink without the skin.

MALTED MILK

Malted milk is a powdered gruel made from a mixture of malted barley, wheat flour, and evaporated whole milk. The powder is used to add its distinctive flavor to beverages and other foods, but it is also used in baking to help the dough cook properly.

Malt powder comes in two forms: diastatic and nondiastatic. Diastatic malt contains enzymes that break down starch into sugar; this is the form bakers add to bread dough to help the dough rise and create a certain crust. Nondiastatic malt, on the other hand, has no active enzymes and is used primarily for flavor, mostly in beverages. It sometimes contains sugar, coloring agents, and other additives, depending on the commercial preparation.

London pharmacist James Horlick developed ideas for an improved, wheat and malt-based nutritional supplement for infants. Despairing of his opportunities in the United Kingdom, Horlick joined his brother William, who had gone to Racine, Wisconsin in the United States, to work at a relative's quarry. In 1873, the brothers formed J & W Horlicks to manufacture their brand of infant food in nearby Chicago. Ten years later, they earned a patent for a new formula enhanced with dried milk. The company originally marketed its new product as "Diastoid", but trademarked the name "malted milk" in 1887.

Despite its origins as a health food for infants and invalids, malted milk found unexpected markets. Explorers appreciated its lightweight, nonperishable, nourishing qualities, and they took malted milk on treks worldwide. William Horlick became a patron of Antarctic exploration, and Admiral Richard E. Byrd named Horlick Mountains, a mountain range in Antarctica, after him. Back in the US, people began drinking Horlick's new beverage for enjoyment. James Horlick returned to England to import his American-made product and was eventually made a baronet. Malted milk became a standard offering at soda shops, and found greater popularity when mixed with ice cream in a "malt", for which malt shops were named.

Uses

- Malted milk biscuits.

- Malted milkshakes.

- Malted hot drinks, such as Horlicks and Ovaltine.

- Malted milk balls: malted milk is used in the popular candy confections Whoppers (manufactured by Hershey Co.), Mighty Malts (manufactured by Necco), and Maltesers (manufactured by Mars, Inc).

- Malted milk is used in some bagel recipes as a substitute for non-diastatic malt powder.

CONDENSED MILK

Condensed milk from a plastic tube package.

Condensed milk is cow's milk from which water has been removed (roughly 60% of it). It is most often found with sugar added, in the form of sweetened condensed milk (SCM), to the extent that the terms "condensed milk" and "sweetened condensed milk" are often used interchangeably today. SCM is a very thick, sweet product, which when canned can last for years without refrigeration if not opened. The product is used in numerous dessert dishes in many countries.

A related product is evaporated milk, which has undergone a lengthier preservation process because it is not sweetened. Evaporated milk is known in some countries as unsweetened condensed milk. Both products have a similar amount of water removed.

A monument to tinned condensed milk at a local
milk-processing factory in Rahachow, Belarus.

According to the writings of Marco Polo, in the thirteenth century the Tatars were able to condense milk. Marco Polo reported that ten pounds (4.5 kg) of milk paste was carried by each man, who would subsequently mix the product with water. However, this probably refers to the soft Tatar curd (katyk), which can be made into a drink (*ayran*) by diluting it, and therefore refers to fermented, not fresh, milk concentrate.

Nicolas Appert condensed milk in France in 1820, and Gail Borden, Jr., in the United States in 1853, in reaction to difficulties in storing milk for more than a few hours. Before this development, milk could be kept fresh for only a short while and was available only in the immediate vicinity of a lactating cow. While returning from a trip to England in 1851, Borden was devastated by the deaths of several children, apparently from poor milk obtained from shipboard cows. With less than a year of schooling and following a series of failures, both of his own and of others, Borden was inspired by the vacuum pan he had seen being used by Shakers to condense fruit juice and managed to reduce milk without scorching or curdling it. Even then his first two factories failed and only the third, built with new partner Jeremiah Milbank in Wassaic, New York, produced a usable milk derivative that was long-lasting and needed no refrigeration.

Probably of equal importance for the future of milk production were Borden's requirements (the "Dairyman's Ten Commandments") for farmers who wanted to sell him raw milk: they were required to wash the cow's udders before milking, keep barns swept clean, and scald and dry their strainers morning and night. By 1858, Borden's milk, sold as Eagle Brand, had gained a reputation for purity, durability and economy.

In 1864, Gail Borden's New York Condensed Milk Company constructed the New York Milk Condensery in Brewster, New York. This was the largest and most advanced milk factory of its day and was Borden's first commercially successful plant. More than 200 dairy farmers supplied 20,000 gallons (76,000 litres) of milk daily to the Brewster plant as demand increased driven by the American Civil War.

The U.S. government ordered huge amounts of condensed milk as a field ration for Union soldiers during the war. This was an extraordinary field ration for the nineteenth century: a typical 10-oz (300-ml) can contained 1,300 calories (5440 kJ), 1 oz (28 g) each of protein and fat, and more than 7 oz (200 g) of carbohydrate.

Soldiers returning home from the war soon spread the word, and by the late 1860s condensed milk was a major product. The first Canadian condensery was built at Truro, Nova Scotia, in 1871. In 1899, E. B. Stuart opened the first Pacific Coast Condensed Milk Company (later known as the Carnation Milk Products Company) plant in Kent, Washington. The condensed milk market developed into a bubble, with too many manufacturers chasing too little demand. In 1911, Nestlé constructed the world's largest condensed milk plant in Dennington, Victoria, Australia. By 1912, high stocks of condensed milk led to a drop in price and many condenseries went out of business.

In 1914, Otto F. Hunziker, head of Purdue University's dairy department, self-published Condensed Milk and Milk Powder: Prepared for the Use of Milk Condenseries, Dairy Students and Pure Food Departments. This text, along with the additional work of Hunziker and others involved with the American Dairy Science Association, standardized and improved condensery operations in the United States and internationally. Hunziker's book was republished in a seventh edition in October 2007 by Cartwright Press.

The First World War regenerated interest in, and the market for, condensed milk, primarily due to its storage and transportation benefits. In the U.S. the higher price for raw milk paid by condenseries created significant problems for the cheese industry.

Production

Raw milk is clarified and standardised to a desired fat to solid-not-fat (SNF) ratio, and is then heated to 85–90 °C (185–194 °F) for several seconds. This heating process destroys some microorganisms, decreases fat separation and inhibits oxidation. Some water is evaporated from the milk and sugar is added until a 9:11 (nearly half) ratio of sugar to (evaporated) milk is reached. The sugar extends the shelf life of sweetened condensed milk. Sucrose increases the liquid's osmotic pressure, which prevents microorganism growth. The sweetened evaporated milk is cooled and lactose crystallization is induced.

Current Use

Condensed milk boiled for several hours to become homemade *dulce de leche*.

Café bombón is a Spanish variation of coffee prepared with condensed milk.

Condensed milk is used in recipes for the Brazilian candy *brigadeiro* (where condensed milk is the main ingredient), key lime pie, caramel candies, and other desserts. Condensed milk and sweetened condensed milk is also sometimes used in combination with clotted cream to make fudge in certain countries such as the United Kingdom.

In parts of Asia and Europe, sweetened condensed milk is the preferred milk to be added to coffee or tea. Many countries in Southeast Asia, such as Vietnam and Cambodia, use condensed milk to flavor their hot or iced coffee. In Malaysia and Singapore, *teh tarik* is made from tea mixed with condensed milk, and condensed milk is an integral element in Hong Kong tea culture. In the Canary Islands, it is served as the bottom stripe in a glass of the local café cortado and, in Valencia, it is served as a café bombón.

Borden's Eagle Brand sweetened condensed milk has noted that ice cream could be made quite simply at home with their product, cream, and various simple flavorings, being ready to serve after as little as four hours.

In New Orleans, sweetened condensed milk is commonly used as a topping on chocolate or similarly cream-flavored snowballs. In Scotland, it is mixed with sugar and butter then boiled to form a popular sweet candy called tablet or Swiss milk tablet, this recipe being very similar to another version of the Brazilian candy brigadeiro called branquinho. In some parts of the Southern United States, condensed milk is a key ingredient in lemon ice box pie, a sort of cream pie. In the Philippines, condensed milk is mixed with some evaporated milk and eggs, spooned into shallow metal containers over liquid caramelized sugar, and then steamed to make a stiffer and more filling version of crème caramel known as leche flan, also common in Brazil under the name pudim de leite.

In Mexico, sweetened condensed milk is one of the main ingredients of the cold cake dessert (the leading brand is "La Lechera", the local version of Swiss Milchmädchen by Nestlé), combined with evaporated milk, marie biscuits, lemon juice, and tropical fruit. In Brazil, this recipe is also done exchanging fruit for puddings, most commonly vanilla and chocolate, known as pave or torta de bolacha. It is also used to make homemade dulce de leche by baking it in an oven. In Brazil, this is done by baking the unopened can in a bain-marie, the result being doce de leite. In Britain and Ireland, the contents of a boiled can is used as the layer between biscuit base and the banana and cream level in banoffee. In Latin American and Central American countries, condensed milk (along with evaporated milk and whole milk or canned cream) is used as a key ingredient in the tres leches cake dessert.

In Poland, it was once common to boil a can of condensed milk in water for about three hours. The resulting product a sweet semi-liquid substance which can be used as a cake icing or put between dry wafers essentially the same as dulce de leche, is called kajmak (although the original kaymak is a product similar to clotted cream). Some manufacturers of condensed milk introduced canned, ready-made kajmak which now is widely commercially produced, and is a national favourite for the sweets fillings. In Russia, the same product is called varionaya sguschyonka (translates as "boiled condensed milk").

Substitutions

Condensed milk can be made from evaporated milk by mixing one measure of evaporated milk with one and a quarter measures of sugar in a saucepan, then heating and stirring the mixture until the sugar is completely dissolved, then cooling. It can also be made by simmering regular milk and sugar, until it is reduced by 60%.

Evaporated Milk

Evaporated milk, known in some countries as "unsweetened condensed milk", is a shelf-stable canned cow's milk product where about 60% of the water has been removed from fresh milk. It differs from sweetened condensed milk, which contains added sugar. Sweetened condensed milk requires less processing to preserve since the added sugar inhibits bacterial growth. The production process involves the evaporation of 60% of the water from the milk, followed by homogenization, canning, and heat-sterilization.

Evaporated milk takes up half the space of its nutritional equivalent in fresh milk. When the liquid product is mixed with a proportionate amount of water (150%), evaporated milk becomes the rough equivalent of fresh milk. This makes evaporated milk attractive for some purposes as it can have a shelf life of months or even years, depending upon the fat and sugar content. This made evaporated milk very popular before refrigeration as a safe and reliable substitute for perishable fresh milk, as it could be shipped easily to locations lacking the means of safe milk production or storage.

Can for Borden's evaporated milk from the second half of the 20th century.
From the Museo del Objeto del Objeto collection in Mexico City.

Evaporated milk is fresh, homogenized milk from which 60 percent of the water has been removed. After the water has been removed, the product is chilled, stabilized, packaged and sterilized. It is commercially sterilized at 240-245 °F (115-118 °C) for 15 minutes. A slightly caramelized flavor results from the high heat process (Maillard reaction), and it is slightly darker in color than fresh milk. The evaporation process concentrates the nutrients and the food energy (kcal); unreconstituted evaporated milk contains more nutrients and calories than fresh milk per unit volume.

Evaporated Milk Infant Formulas

Colwell & Brothers cast iron vacuum
pan, for evaporating milk.

In the 1920s and 1930s, evaporated milk began to be widely commercially available at low prices. The Christian Diehl Brewery, for instance, entered the business in 1922, producing Jerzee brand evaporated milk as a response to the Volstead Act. Several clinical studies from that time period suggested that babies fed evaporated milk formula thrived as well as breastfed babies. Modern guidelines from the World Health Organization consider breastfeeding, in most cases, to be healthier for the infant because of the colostrum in early milk production, as well as the specific nutritional content of human breast milk.

SCALDED MILK

Scalded milk is dairy milk that has been heated to 83 °C (181 °F). At this temperature, bacteria are killed, enzymes in the milk are destroyed, and many of the proteins are denatured. Since most milk sold today is pasteurized, which accomplishes the first two goals, milk is typically scalded to increase its temperature, or to change the consistency or other cooking interactions due to the denaturing of proteins.

During scalding, a milk watcher (a cooking utensil) may be used to prevent both boiling over and scorching (burning) of the milk.

Uses

- Scalded milk is called for in the original recipes for Béchamel sauce, as adding hot liquid, including milk, to a roux was thought less likely to make a lumpy sauce or one tasting of raw flour.

- Scalded and cooled milk is used in bread and other yeast doughs, as pasteurization does not kill all bacteria, and with the wild yeasts that may also be present, these can alter the texture and flavor. Recipes old enough to have been based on hand-milked, slowly cooled, unpasteurized milk specify scalded milk with much more justification, and modern cookbooks tend to maintain the tradition. In addition, scalding milk improves the rise due to inhibition of bread rise by certain undenatured milk proteins.

- Scalded milk is used in yogurt to make the proteins unfold, and to make sure that all organisms that could outcompete the yogurt culture's bacteria are killed. In traditional yogurt making, as done in the Eastern Mediterranean and Near East, the milk is often heated in flat pans until reduced to about half. Whatever the effect of scalding on milk protein may be, it is mainly this concentrating that reduces whey separation. Modern commercial processors use dried or concentrated milk or vegetable gums and gelatins such as pectin, carrageenan, or agar ("vegetable gelatin") to prevent whey separation in yogurt.

- Café au lait, baked milk, and ryazhenka also use scalded milk.

- Scalded and cooled milk is used in many recipes for raised doughnuts, probably for the same reason it is so often specified in bread recipes.

However, latte art does not use scalded milk, as scalding destroys the microfoam texture; milk for latte art is heated to below the scalding point.

SOY MILK

Soy milk, also known as soya milk, is a plant-based drink produced by soaking and grinding soybeans, boiling the mixture, and filtering out remaining particulates. It is a stable emulsion of oil, water, and protein. Its original form is a natural by-product of the manufacture of tofu. It became a common beverage in Europe and North America in the later half of the 20th century, especially as production techniques were developed to give it taste and consistency more closely resembling dairy milk. Along with similar vegetable-based "milks", like almond and rice milk, soy milk may be used as a substitute for dairy milk by individuals who are vegan or lactose intolerant, while others may consume it from environmental concerns.

Bottled soy milk "Vitamilk" in Thailand.

Taste

Soy milk flavor quality differs according to the cultivar of soybean used in its production. Even in China, the desirable sensory qualities are a mouthfeel (smooth but thick), color (off-white), and appearance (creamy) resembling milk. These traits along with a pleasing aroma are positively correlated with a soy milk's content of proteins, soluble solids, and oil. In the United States, testing suggests consumers prefer viscous soy milk with sweet aromatic flavors like vanilla and actively dislike the "beany" or "brothy" flavors resembling traditional doujiang.

Preparation

Soy milk is made from whole soybeans or full-fat soy flour. The dry beans are soaked in water for a minimum of three hours up to overnight depending on the temperature of the water. The rehydrated beans then undergo wet grinding with enough added water to give the desired solids content to the final product which has a protein content of 1–4%, depending on the method of production. The ratio of water to beans on a weight basis is 10:1 for traditional soy milk. The resulting slurry or purée is brought to a boil in order to improve its taste properties by heat inactivating soybean trypsin inhibitor, improve its flavor, and to sterilize the product. Heating at or near the boiling point is continued for a period of time, 15–20 minutes, followed by the removal of insoluble residues (soy pulp fiber) by filtration. Processing requires the use of an anti-foaming agent or natural defoamer during the boiling step. Bringing filtered soy milk to a boil avoids the problem of foaming. It is generally opaque, white or off-white in color, and approximately the same consistency as cow's milk. Quality attributes during preparation include germination time for the beans used, acidity, total protein and carbohydrates, phytic acid content, and viscosity.

Ecological Impact

Using soybeans to make milk instead of raising cows may be ecologically advantageous. Cows require much more energy in order to produce milk, since the farmer must feed the animal, which can consume up to 24 kilograms (53 lb) of food in dry matter basis and 90 to 180 litres (24 to 48 US gal) of water a day, producing an average of 40 kilograms (88 lb) of milk a day. Legumes, including the soybean plant, also replenish the nitrogen content of the soil in which they are grown.

Mean greenhouse gas emissions for one glass (200g) of different milks.	
Milk Types	Greenhouse gas emissions (kg CO_2-Ceq per 200g).
Cow milk	0.62
Rice milk	0.23
Soy milk	0.21
Oat milk	0.19
Almond milk	0.16

The cultivation of soybeans in South America is a cause of deforestation (specifically in the Amazon rainforest) and a range of other large-scale environmental harm. However, the majority of soybean cultivation worldwide, especially in South America where cattle farming is widespread, is intended for livestock fodder rather than soy milk production.

EXTENDED SHELF LIFE MILK

ESL is an acronym and stands for "Extended Shelf Life" which means, as the name implies, that the milk is "longer lasting" and does not spoil as quickly as fresh milk.

Just like fresh milk, ESL milk has to be stored in the fridge regardless of whether it is opened or unopened. This way ESL milk will last drinkable for up to 4 weeks (unopened). In contrast, fresh milk will last 6 – 12 days in the fridge and ultra high temperature (UHT) milk even 3 – 6 months.

Taste and Nutritional Facts

Even though the taste and nutritional composition of the milk are increasingly changed with every additional thermal and mechanical treatment, ESL milk shows good results in terms of preservation of taste and nutrition. In fact, test subjects in various experiments could not identify by taste which product is ESL milk and which traditionally pasteurized milk.

The overall sensorial properties of pasteurized milk are slightly better than the ones of ESL milk. The average customer, however, is not able to determine those differences.

In 2009 the Max Rubner Institute discovered that the concentration of vitamins in ESL milk is almost similar to the concentration of pasteurized milk. Various experiments showed that only 0 – 5 percent of the vitamins get lost during the various process steps.

In many cases also milk fat, milk sugar (lactose) and minerals like calcium have been identical to pasteurized milk.

In terms of protein structure, some of the molecules are broken up in the ESL milk during the thermal treatment just like in the case of UHT milk. But this is rather an advantage than a disadvantage as this means that the milk becomes more digestible for humans. Beside the above facts the nutritional value of ESL milk is basically identical to pasteurized milk.

Market Situation

How and when did the ESL milk enter the market and what is the current market situation?

Everything started (in Germany at least) in the mid-90s when an organic dairy farmer ended up with more organic milk than he could actually sell. However, he did not want his high-quality milk to be processed to UHT milk.

So he approached the biggest milk processor in his region and together they came up with the idea to make the first HT ESL milk. In 1996 the first ESL milk could be bought in German supermarkets.

Even though ESL milk came along with big advantages for farmers and milk processors, the German retail was rather reluctant. So that ESL milk remained a niche product until 2005.

Everything changed after 2005 when there were no strict regulations that prescribe how to advertise and name ESL milk. That means there was no need to label ESL milk as an HT product or notify the customer in any other way that it differs from traditionally pasteurized milk. The reason for this approach by the EU regulators were the numerous production processes which were developed to make ESL milk like e.g. microfiltration, direct and indirect heating etc. As a consequence pasteurized milk was simply replaced with ESL milk, still keeping the label "Fresh Milk".

This fact made the fresh milk market share of ESL milk jump from 14 % in 2005 to as high as 80 % in 2008. Nowadays traditional fresh milk can be considered the niche product that ESL milk was before.

With the rise of the market share, the price of ESL milk dropped considerably. Although ESL Milk had a relatively high price in the early years due to the status of an innovative product, the price today is just as under pressure as that of other milk products.

Methods of Esl Production

The following five methods are currently used to produce ESL milk:

1. Indirect heating through a plate or tubular heat exchangers (PHE/THE).

2. Direct heating through steam infusion or steam injection.

3. Microfiltration.

4. Depth filtration.

5. Double bactofugation.

All the above processing alternatives are equally valid ways to make ESL milk, there are no laws or regulations which prescribe a certain production method for ESL milk.

What follows is a more specific description of all the single methods.

Indirect Heating through a Plate or Tubular Heat Exchanger

This is probably the most common method to produce ESL milk (together with direct heating) because this method allows achieving the highest shelf live results.

For this process traditionally pasteurized (and standardized) milk is needed as a basis. After the milk is pasteurized it is then gradually heated by flowing through a PHE or THE until it reaches a temperature of 127 °C which is then held for up to three seconds. After that, the milk is either filled or stored.

Storage temperature is at around 5 °C and may by no means rise above 8 °C otherwise we would corrupt the shelf life of the milk.

Direct Heating through Steam Infusion or Steam Injection

This is one of the most common ways to produce ESL milk. Just like for indirect heating we need pasteurized and standardized milk as our base product.

The milk is heated in a two-stage process, first to 70 °C – 85 °C and then to 127 °C – 130 °C by means of direct heating with steam. The final milk temperature is held for up to three seconds before it is cooled down to 70 °C – 85 °C again. In this kind of systems cooling is done in a vessel also called a flash cooler which has two purposes:

- Super fast cooling under a specific vacuum.

- Removing the water that was taken on in the form of steam through a rapid drop of temperature.

Lastly, the milk is homogenized with an aseptic homogenizer and cooled down to 5 °C storage temperature.

Microfiltration (MF)

This method was developed to further decrease thermal stress on the milk and to improve its sensory qualities. The first step in this process is to separate the cream from the milk. Afterward, the skimmed milk is sterilized through MF. MF plants apply a so-called cross-flow technique where ceramic membranes with a pore size between 0.8 µm and 1.4 µm are used.

99.5 – 99.9 % of all germs are held back by the membranes and gathered in the MF retentate. The cream that was previously separated from the milk is heated to temperatures of up to 110 °C and held at this level for four to six seconds before added to the filtrated milk to reach the required fat content (skimmed, semi-skimmed, whole).

Cheese dairies usually have an additional process step. They reuse the retentate by adding it to the cream and homogenizing it before adding the cream to the milk for standardization. If you want ESL milk as the final product this is not necessarily recommended to do because it can decrease your shelf live.

As the last step, the standardized milk is pasteurized, homogenized and finally cooled down to storage temperature.

Depth Filtration (DF)

DF is in all its processing steps identical to MF, except for the filtration process itself.

DF uses another filtration principle than MF which is derived from the beverage industry. In DF we have two filter units, one pre-filter, and a final-filter. Both units consist of several polypropylene filter candles with pores that measure 0.3 μm in the pre-filters and 0.2 μm in the final filters.

A retentate is not produced. In this kind of filtration as all the germs and solids which are held back by the filters are collected in the pores and not on their surface.

Double Bactofugation

This is a more recent method to produce ESL milk even though it uses a very traditional approach by simply enhancing milk pasteurization by integrating two bacteria removal separators (bactofuges) in series.

This way a considerable amount of spores are removed from the milk mechanically. As the spores have a higher specific density than the milk and the cream they can be separated by using centrifugal force.

Of course, the milk also goes through traditional pasteurization after double bactofugation and separation (milk/cream). The shelf life of this milk can reach 20 days or more.

COMPOSITION OF MILK

The milk composition describes the chemical and physical properties and effects of pasteurization on the compounds in milk. position is provided below:

In general, the gross composition of cow's milk in the U.S. is 87.7% water, 4.9% lactose (carbohydrate), 3.4% fat, 3.3% protein, and 0.7% minerals (referred to as ash). Milk composition varies depending on the species (cow, goat, sheep), breed (Holstein, Jersey), the animal's feed, and the stage of lactation. Although there are minor variations in milk composition, the milk from different cows is stored together in bulk tanks and provides a relatively consistent composition of milk year round in the U.S.

Milk composition and yield can also be influenced by other factors, including age and body weight at kidding, gestation and dry periods, body condition at kidding, plane of nutrition, etc. In terms of plane of nutrition, increasing the energy intake increases the level of milk production toward the goat's inherited potential. Factors that increase milk yield in cows or goats are increased body weight, advancing age, increased plane of nutrition, fall and winter kidding, moderate or cool environmental temperatures, and good body condition at kidding.

Milk composition varies among phylogenetic taxa. These compositional differences are thought to be a result of both genetic and environmental variation that respond to different selective pressures.

Intergenus Variation in Milk Composition

Although the etiological basis (including selective factors, if they exist) for evolutionary changes that have clearly taken place in terms of milk production and composition are not well understood, examples of variation in the evolution of lactation have been well described and explored across mammalian genera, particularly for total lipid content, the primary source of energy during early life. For instance, several reports related to milk lipids in bears, true seals, and whales: large mammals that fast during much or all of their lactation. These mammals produce milk with exceptionally high lipid content (200–500 g/L) while fasting by mobilizing up to 40% of their body weight. These species have greater adipose and protein stores relative to rates of milk production, so that this intensive lactation period can occur without jeopardizing maternal health and subsequent reproductive efforts. This is not the case for other taxa, for example, H. sapiens; human milk typically contains 30–50 g/L lipids. As described by Capuco and Akers, these and other differences in lactation strategy and milk production are thought to have evolved to meet the diverse reproductive strategies and environmental demands of different mammals.

Concentrations and profiles of milk carbohydrates also vastly differ among genera. For instance, whereas primates and some ruminants produce milk with relatively high levels of lactose (~50–70 g/L), rodent milk has less (~30–50 g/L), and that of pinnipeds, cetaceans, bears, and marsupials often even less. Total complex carbohydrate (oligosaccharide) content and composition also vary among genera. For instance, bovine milk typically has a much lower oligosaccharide concentration and diversity than human milk (0.05 and 5–15 g/L, respectively), and the majority of the bovine oligosaccharides are more simplistic in their chemical structure than are those found in human milk. To date, only 40 different bovine milk oligosaccharides have been identified in the colostrum of Holstein-Friesian cows, whereas there are reportedly at least 100 human milk oligosaccharides (HMOs). Lactose concentration and oligosaccharide content and diversity also differ greatly among New and Old World monkeys, and between humans and apes. As proposed by Urashima et al., it is possible that these vast differences in lactose and oligosaccharide profiles in milk are a result of gradual and bifurcated shifts in the function of the evolving mammary gland—first acting primarily as an innate immune tissue with high expression of lysozyme (an enzyme that helps destroy bacterial cell walls) and eventually, as the DNA encoding for α-lactalbumin evolved from DNA encoding for lysozyme, to a nutritive organ. Indeed, milk oligosaccharides predominate over lactose in monotremes and marsupials, whereas milk produced by most of Eutherian taxa contains lactose as the dominant carbohydrate. Urashima et al. have speculated that the high oligosaccharide concentration and somewhat unique oligosaccharide profile of human milk (compared with closely related apes) is an evolutionary consequence of human neonates being relatively more altricial and needing additional milk-borne protection from pathogenic organisms.

Milk also differs in micronutrient concentrations. For instance, iron concentrations are low in human and bovine milk (0.3–0.9 and 0.2–0.6 µg/mL, respectively) relative to that produced by dogs (6.5–7.6 µg/mL). Iron, although essential for growth and development throughout the human lifecycle and still a major public health concern globally, is also necessary for the growth and replication of many bacteria. Iron's role in infectious disease (e.g., malaria) etiology is well known. As such, the maintenance of very low iron levels in human milk has been hypothesized by some to represent an important evolutionary tradeoff favoring reduced risk of infectious disease during infancy over increased risk for mild iron deficiency. This begs the questions of whether all human infants (regardless of their infectious disease risk) benefit from similar iron intake levels and if current practices related to iron supplementation during infancy and fortification of infant formulas with relatively high levels of iron are wise across the board.

There exist ample data suggesting that milk composition, at both macronutrient and micronutrient levels, varies among genera and that these differences may have evolved to maintain health across various environmental constraints (e.g., microbial exposures and risks) and behavioral contexts.

Intraspecies Variation in Milk Composition

Certain milk components can also vary greatly even within a species. Take, for example, human milk lipids (including their fatty acid profiles), which are affected by maternal diet and maternal body composition. One of the first studies to provide evidence of this was conducted by Insull et al, who reported rapid, reversible shifts in milk fat in response to dietary macronutrient manipulation in a single, hospitalized breastfeeding woman. Karmarkar et al. also observed a 25% increase in milk fat content with dietary lipid supplementation. Similarly, Park et al. showed that milk fat was 18% higher when subjects consumed a high–dairy fat diet, and Mohammad et al. reported 12% greater milk fat when women consumed a high-lipid diet compared with an isocaloric, low-fat diet. Martín et al. provide similar data for forager-horticulturalists in Bolivia.

In addition to acute (proximal) effects of diet, long-term energy balance is also associated with total milk fat content: women with higher maternal body mass or body mass index (BMI) tend to produce higher-fat milk than those with lower BMI. This difference may both protect women living in regions with chronic malnutrition from mobilizing excess body weight during lactation and result in smaller offspring who will, across their entire lifespan, require fewer calories—an example of the "small but healthy" hypothesis. However, as discussed by Prendergast and Humphrey, the tradeoff for stunted growth and decreased caloric requirements across the lifespan is likely increased morbidity and mortality in early life.

It is noteworthy that whereas lipids and selected other components (e.g., B vitamins) tend to be variable in milk produced by different women, concentrations of other components (e.g., iron) appear to be more conserved. As described previously for differences in milk iron concentrations among genera, it is thought that there may exist adaptive advantages to maintaining low iron content in human milk. This low level of variation for certain milk constituents, such as iron, may indicate a protective advantage that is not modified by other more fluid factors such as maternal nutritional status, environmental microbial exposures, and feeding practices. Conversely, greater plasticity in some components (e.g., lipids) might represent an important ability to adjust milk production and/or composition to adapt to more acute environmental conditions.

In addition to the essential and energy-yielding nutrients that are in milk, such as iron and lipids, there is also an impressive host of other biologically active components whose concentrations vary (or are conserved) among women. Of particular interest to us in this regard is the relatively high abundance and variability in complex carbohydrates and microbial taxa in milk.

MILK CARBOHYDRATE LACTOSE

Carbohydrates are made up of molecules called saccharides. Simple saccharides contain 1 or 2 molecules and are called monosaccharides or disaccharides, or, more commonly, sugars. Oligosaccharides and polysaccharides are chains that contain a few to many sugar molecules and may be referred to as starches.

The monosaccharides important in food and health are glucose (sometimes called dextrose), fructose, and galactose. The disaccharides are sucrose (glucose + fructose), lactose (glucose + galactose), and maltose (glucose + glucose). The 2 sugar molecules in disaccharides are bonded together and this bond must be broken before the sugars can be used by the body for energy and other body functions. Starches are long chains of glucose that can be straight or have branches, and there are several ways in which the molecules within starches can bond to each other. Starches from different sources, such as wheat or corn, have unique functional properties. The structure of starches can be modified to improve their functional properties and increase their use in foods.

Milk Carbohydrate Chemistry

Milk contains approximately 4.9% carbohydrate that is predominately lactose with trace amounts of monosaccharides and oligosaccharides. Lactose is a disaccharide of glucose and galactose. The structure of lactose is:

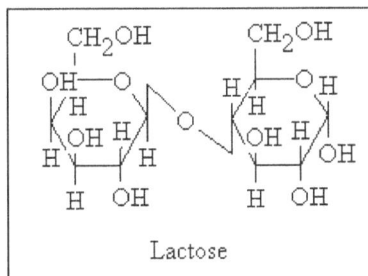

Lactose

Lactose Physical Properties

Lactose is dissolved in the serum (whey) phase of fluid milk. Lactose dissolved in solution is found in 2 forms, called the α-anomer and ß-anomer, that can convert back and forth between each other. The solubility of the 2 anomers is temperature dependent and therefore the equilibrium concentration of the 2 forms will be different at different temperatures. At room temperature (70 °F, 20 °C) the equilibrium ratio is approximately 37% α- and 63% ß-lactose. At temperatures above 200 °F (93.5 °C) the ß-anomer is less soluble so there is a higher ratio of α- to ß-lactose. The type of anomer present does not affect the nutritional properties of lactose.

Lactose crystallization occurs when the concentration of lactose exceeds its solubility. The physical properties of lactose crystals are dependent on the crystal type and can greatly influence their use in foods. Temperature affects the equilibrium ratio of the α- and ß-lactose anomers, as described above. Lactose crystals formed at temperatures below 70 °F (20 °C) are mainly α-lactose crystals. The α-monohydrate lactose crystals are very hard and form, for example, when ice cream goes through numerous warming and freezing cycles. This results in an undesirable gritty, sandy texture in the ice cream. Gums are often used in ice cream to inhibit lactose crystallization. The crystal form of ß-lactose is sweeter and more soluble than the α-monohydrate lactose and may be preferred in some bakery applications. When a lactose solution is rapidly dried it does not have time to crystallize and forms a type of glass. Lactose glass exists in milk powders and causes clumping. The clumping is desirable because it results in a milk powder that dissolves instantly in water.

Influence of Heat Treatments on Lactose Properties

The normal pasteurization conditions used for fluid milk have no significant effect on lactose. The higher temperatures used for ultra high temperature (UHT) pasteurization of extended shelf life products and spray drying can cause browning and isomerization reactions, which may affect product quality and nutritional properties. The browning reaction, called the Maillard reaction, occurs between the lactose and protein in milk and produces undesirable flavors and color, and decreases the available content of the amino acid lysine in milk protein. The isomerization reaction is a molecular rearrangement of lactose to lactulose. Lactulose is produced for use by the pharmaceutical industry in pill production.

VITAMINS AND MINERALS IN MILK

Vitamins have many roles in the body, including metabolism co-factors, oxygen transport and antioxidants. They help the body use carbohydrates, protein, and fat.

Milk contains the water soluble vitamins thiamin (vitamin B1), riboflavin (vitamin B2), niacin (vitamin B3), pantothenic acid (vitamin B5), vitamin B6 (pyridoxine), vitamin B12 (cobalamin), vitamin C, and folate. Milk is a good source of thiamin, riboflav in and vitamin B12 . Milk contains small amounts of niacin, pantothenic acid, vitamin B6, vitamin C, and folate and is not considered a major source of these vitamins in the diet.

Milk contains the fat soluble vitamins A, D, E, and K. The content level of fat soluble vitamins in dairy products depends on the fat content of the product. Reduced fat (2% fat), lowfat (1% fat), and skim milk must be fortified with vitamin A to be nutritionally equivalent to whole milk. Fortification of all milk with vitamin D is voluntary. Milk contains small amounts of vitamins E and K and is not considered a major source of these vitamins in the diet.

Minerals in Milk

Minerals have many roles in the body including enzyme functions, bone formation, water balance maintenance, and oxygen transport.

Milk is a good source of calcium, magnesium, phosphorus, potassium, selenium, and zinc. Many minerals in milk are associated together in the form of salts, such as calcium phosphate. In milk approximately 67% of the calcium, 35% of the magnesium, and 44% of the phosphate are salts bound within the casein micelle and the remainder are soluble in the serum phase. The fact that calcium and phosphate are associated as salts bound with the protein does not affect the nutritional availability of either calcium or phosphate.

Milk contains small amounts of copper, iron, manganese, and sodium and is not considered a major source of these minerals in the diet.

Effects of Heat Treatments and Light Exposure on the Vitamin and Mineral Content in Milk

The mild heat treatment used in the typical high temperature short time (HTST) pasteurization of fluid milk does not appreciably affect the vitamin content. However, the higher heat treatment used in ultra high temperature (UHT) pasteurization for extended shelf combined with the increased storage life of these products does cause losses of some water-soluble vitamins. Thiamin is reduced from 0.45 to 0.42 mg/L, vitamin B 12 is reduced from 3.0 to 2.7 µg/L, and vitamin C is reduced from 2.0 to 1.8 mg/L (Potter et al., 1984). Riboflavin is a heat stable vitamin and is not affected by severe heat treatments.

Calcium phosphate will migrate in and out of the casein micelle with changes in temperature. This process is reversible at moderate temperatures. This does not affect the nutritional properties of milk minerals. At very high temperatures the calcium phosphate may precipitate out of solution which causes irreversible changes in the casein micelle structure.

Exposure to light will decrease the riboflavin and vitamin A content in milk. Milk should be stored in containers that provide barriers to light (opaque plastic or paperboard) to maximize vitamin retention.

MILK FAT

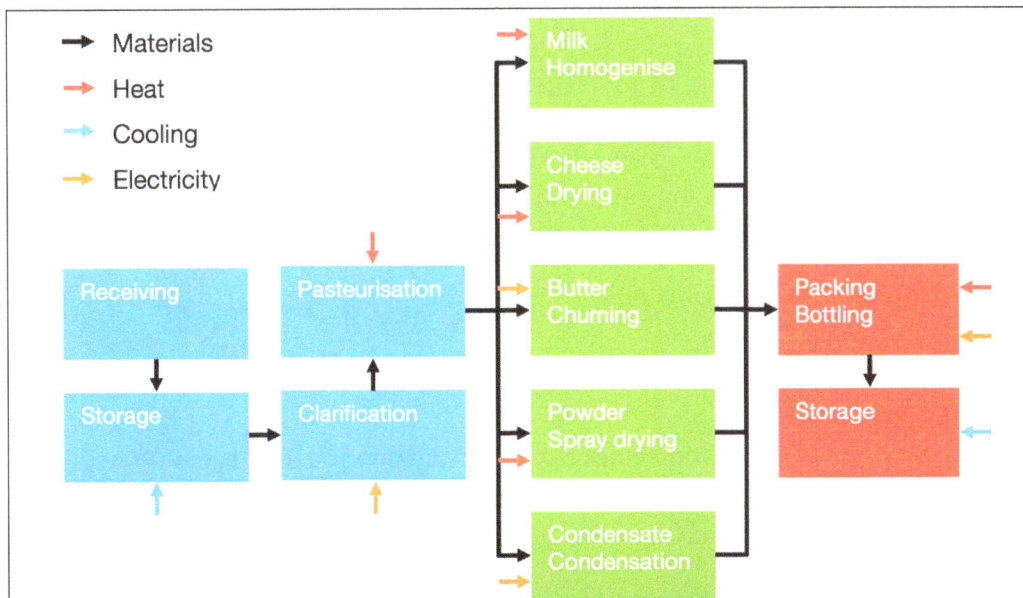

Chart of milk products and production relationships, including milk.

The fat content of milk is the proportion of milk, by weight, made up by butterfat. The fat content, particularly of cow's milk, is modified to make a variety of products. The fat content of milk is usually stated on the container, and the color of the label or milk bottle top varied to enable quick recognition.

Health and Nutrition

Fat has more nutritional energy per cup, but researchers found that in general low fat milk drinkers do absorb less fat, and will compensate for the energy deficit by eating more carbohydrates. They also found that the lower milk fat drinkers also ate more fruits and vegetables, while the higher milk fat drinkers also ate more meat and sweets.

Nutrition intake between whole milk drinkers and skimmed or low fat drinkers is different. An analysis of a survey done by the U. S. Department of Agriculture showed that consumers of reduced or low fat milk had greater intake of vitamins, minerals, and dietary fiber compared to the group of whole milk drinkers, yet zinc, vitamin E, and calcium were all under consumed in each group. The conclusion was that the whole milk drinkers were more likely to choose foods that were less micronutrient-dense, which resulted in their less healthful diets.

Researchers, found that drinking full-fat milk may actually be better for your heart than drinking skimmed milk. This is because it boosted levels of 'good' HDL cholesterol in the bloodstream.

Methods for Reducing Fat Content

To reduce the fat content of milk, e.g. for skimmed or semi-skimmed milk, all of the fat is removed and then the required quantity returned. The fat content of the milk produced by cows can also be altered, by selective breeding and genetic modification. For example, scientists in New Zealand have bred cows that produce skimmed milk (less than 1% fat content).

Methods of Detecting Fat Content

Milk's fat content can be determined by experimental means, such as the Babcock test or Gerber Method. Before the Babcock test was created, dishonest milk dealers could adulterate milk to falsely indicate a higher fat content. In 1911, the American Dairy Science Association's Committee on Official Methods of Testing Milk and Cream for Butterfat met in Washington DC with the U.S. Bureau of Dairying, the U.S. Bureau of Standards and manufacturers of glassware. Standard specifications for the Babcock methodology and equipment were published as a result of this meeting. Improvements to the Babcock test have continued.

MILK PROTEIN

Proteins are chains of amino acid molecules connected by peptide bonds.

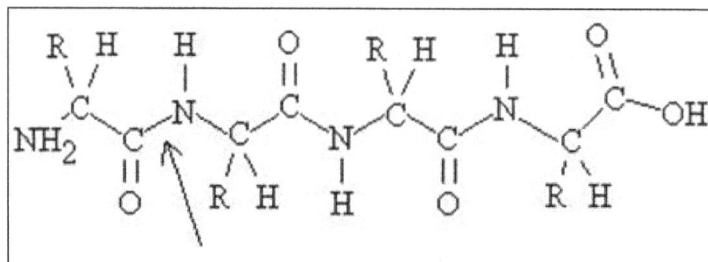

There are many types of proteins and each has its own amino acid sequence (typically containing hundreds of amino acids). There are 22 different amino acids that can be combined to form protein chains. There are 9 amino acids that the human body cannot make and must be obtained from the diet. These are called the essential amino acids.

The amino acids within protein chains can bond across the chain and fold to form 3-dimensional structures. Proteins can be relatively straight or form tightly compacted globules or be somewhere in between. The term "denatured" is used when proteins unfold from their native chain or globular shape. Denaturing proteins is beneficial in some instances, such as allowing easy access to the protein chain by enzymes for digestion, or for increasing the ability of the whey proteins to bind water and provide a desirable texture in yogurt production.

Milk Protein Chemistry

Milk contains 3.3% total protein. Milk proteins contain all 9 essential amino acids required by humans. Milk proteins are synthesized in the mammary gland, but 60% of the amino acids used to build the proteins are obtained from the cow's diet. Total milk protein content and amino acid composition varies with cow breed and individual animal genetics.

There are 2 major categories of milk protein that are broadly defined by their chemical composition and physical properties. The casein family contains phosphorus and will coagulate or precipitate at pH 4.6. The serum (whey) proteins do not contain phosphorus, and these proteins remain in solution in milk at pH 4.6. The principle of coagulation, or curd formation, at reduced pH is the basis for cheese curd formation. In cow's milk, approximately 82% of milk protein is casein and the remaining 18% is serum, or whey protein.

The casein family of protein consists of several types of caseins (α-s1, α-s2 , ß, and 6) and each has its own amino acid composition, genetic variations, and functional properties. The caseins have a relatively random, open structure due to the amino acid composition (high proline content). The high phosphate content of the casein family allows it to associate with calcium and form calcium phosphate salts. The abundance of phosphate allows milk to contain much more calcium than would be possible if all the calcium were dissolved in solution, thus casein proteins provide a good source of calcium for milk consumers. The 6-casein is made of a carbohydrate portion attached to the protein chain and is located near the outside surface of the casein micelle. In cheese manufacture, the 6-casein is cleaved between certain amino acids, and this results in a protein fragment that does not contain the amino acid phenylalanine. This fragment is called milk glycomacropeptide and is a unique source of protein for people with phenylketonuria.

The serum (whey) protein family consists of approximately 50% ß-lactoglobulin, 20% α-lactalbumin, blood serum albumin, immunoglobulins, lactoferrin, transferrin, and many minor proteins and enzymes. Like the other major milk components, each whey protein has its own characteristic composition and variations. Whey proteins do not contain phosphorus, by definition, but do contain a large amount of sulfur-containing amino acids. These form disulfide bonds within the protein causing the chain to form a compact spherical shape. The disulfide bonds can be broken, leading to loss of compact structure, a process called denaturing. Denaturation is an advantage in yogurt production because it increases the amount of water that the proteins can bind, which improves the texture of yogurt. This principle is also used to create specialized whey protein ingredients with unique functional properties for use in foods. One example is the use of whey proteins to bind water in meat and sausage products.

The functions of many whey proteins are not clearly defined, and they may not have a specific function in milk but may be an artifact of milk synthesis. The function of ß-lactoglobulin is thought to be a carrier of vitamin A. It is interesting to note that ß-lactoglobulin is not present in human milk. α-Lactalbumin plays a critical role in the synthesis of lactose in the mammary gland. Immunoglobulins play a role in the animal's immune system, but it is unknown if these functions are transferred to humans. Lactoferrin and transferrin play an important role in iron absorption and there is interest in using bovine milk as a commercial source of lactoferrin.

Milk Protein Physical Properties

The caseins in milk form complexes called micelles that are dispersed in the water phase of milk. The casein micelles consist of subunits of the different caseins (α-s1, α-s2 and ß) held together by calcium phosphate bridges on the inside, surrounded by a layer of 6-casein which helps to stabilize the micelle in solution.

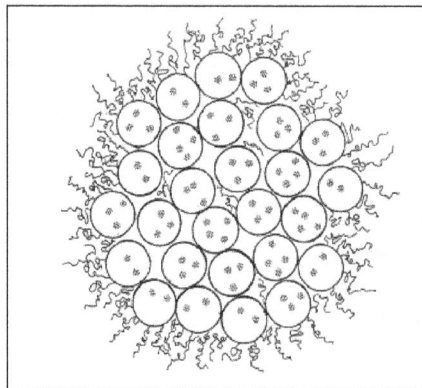

Casein micelles are spherical and are 0.04 to 0.3 µm in diameter, much smaller than fat globules which are approximately 1 µm in homogenized milk. The casein micelles are porous structures that allow the water phase to move freely in and out of the micelle. Casein micelles are stable but dynamic structures that do not settle out of solution. They can be heated to boiling or cooled, and they can be dried and reconstituted without adverse effects. ß-casein, along with some calcium phosphate, will migrate in and out of the micelle with changes in temperature, but this does not affect the nutritional properties of the protein and minerals.

The whey proteins exist as individual units dissolved in the water phase of milk.

Deterioration of Milk Protein

Proteins can be degraded by enzyme action or by exposure to light. The predominant cause of protein degradation is through enzymes called proteases. Milk proteases come from several sources: the native milk, airborne bacterial contamination, bacteria that are added intentionally for fermentation, or somatic cells present in milk. The action of proteases can be desirable, as in the case of yogurt and cheese manufacture, so, for these processes, bacteria with desirable proteolytic properties are added to the milk. Undesirable degradation (proteolysis) results in milk with off-flavors and poor quality. The most important protease in milk for cheese manufacturing is plasmin because it causes proteolysis during ripening which leads to desirable flavors and texture in cheese.

Two amino acids in milk, methionine and cystine are sensitive to light and may be degraded with exposure to light. This results in an off-flavor in the milk and loss of nutritional quality for these 2 amino acids.

Influence of Heat Treatment on Milk Proteins

The caseins are stable to heat treatment. Typical high temperature short time (HTST) pasteurization conditions will not affect the functional and nutritional properties of the casein proteins. High temperature treatments can cause interactions between casein and whey proteins that affect the functional but not the nutritional properties. For example, at high temperatures, ß-lactoglobulin can form a layer over the casein micelle that prevents curd formation in cheese.

The whey proteins are more sensitive to heat than the caseins. HTST pasteurization will not affect the nutritional and functional properties of the whey proteins. Higher heat treatments may cause denaturation of ß-lactoglobulin, which is an advantage in the production of some foods (yogurt) and ingredients because of the ability of the proteins to bind more water. Denaturation causes a change in the physical structure of proteins, but generally does not affect the amino acid composition and thus the nutritional properties. Severe heat treatments such as ultra high pasteurization may cause some damage to heat sensitive amino acids and slightly decrease the nutritional content of the milk. The whey protein α-lactalbumin, however, is very heat stable.

MILK LIPIDS

The fat content of milk is of economic importance because milk is sold on the basis of fat. Milk fatty acids originate either from microbial activity in the rumen, and transported to the secretory cells via the blood and lymph, or from synthesis in the secretory cells. The main milk lipids are a class called triglycerides which are comprised of a glycerol backbone binding up to three different fatty acids. The fatty acids are composed of a hydrocarbon chain and a carboxyl group. The major fatty acids found in milk are:

Long Chain

- C14 - myristic 11%
- C16 - palmitic 26%
- C18 - stearic 10%
- C18:1 - oleic 20%

Short Chain (11%)

- C4 - butyric

- C6 - caproic

- C8 - caprylic

- C10 – capric

Butyric fatty acid is specific for milk fat of ruminant animals and is responsible for the rancid flavour when it is cleaved from glycerol by lipase action.

Saturated fatty acids (no double bonds), such as myristic, palmitic, and stearic make up two thirds of milk fatty acids. Oleic acid is the most abundant unsaturated fatty acid in milk with one double bond. While the cis form of geometric isomer is the most common found in nature, approximately 5% of all unsaturated bonds are in the trans position as a result of rumen hydrogenation.

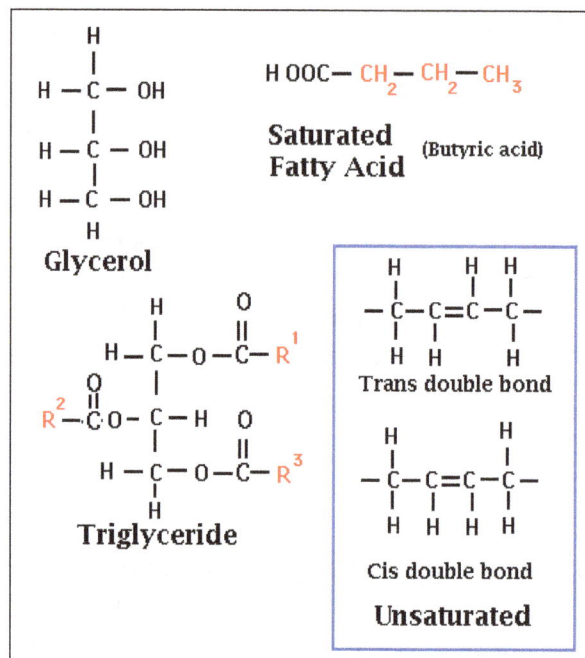

Triglycerides account for 98.3% of milkfat. The distribution of fatty acids on the triglyceride chain, while there are hundreds of different combinations, is not random. The fatty acid pattern is important when determining the physical properties of the lipids. In general, the SN1 position binds mostly longer carbon length fatty acids, and the SN3 position binds mostly shorter carbon length and unsaturated fatty acids. For example:

- C4 - 97% in SN3

- C6 - 84% in SN3

- C18 - 58% in SN1

The small amounts of mono- , diglycerides, and free fatty acids in fresh milk may be a product of early lipolysis or simply incomplete synthesis. Other classes of lipids include phospholipids (0.8%) which are mainly associated with the fat globule membrane, and cholesterol (0.3%) which is mostly located in the fat globule core.

The physical properties of milkfat can be summerized as follows:

- Density at 20 °C is 915 kg m.

- Refractive index (589 nm) is 1.462 which decreases with increasing temperature.

- Solubility of water in fat is 0.14% (w/w) at 20 °C and increases with increasing temperature.

- Thermal conductivity is about 0.17 J m(-1) s(-1) K(-1) at 20 °C.

- Specific heat at 40 °C is about 2.1kJ kg(-1) K(-1).

- Electrical conductivity is <10(-12) ohm(-1) cm(-1).

- Dielectric constant is about 3.1.

The brackets around numbers denote superscript.

At room temperature, the lipids are solid, therefore, are correctly referred to as "fat" as opposed to "oil" which is liquid at room temperature. The melting points of individual triglycerides ranges from -75 °C for tributyric glycerol to 72 °C for tristearin. However, the final melting point of milkfat is at 37 °C because higher melting triglycerides dissolve in the liquid fat. This temperature is significant because 37 °C is the body temperature of the cow and the milk would need to be liquid at this temperature. The melting curves of milkfat are complicated by the diverse lipid composition:

- Trans unsaturation increases melting points.

- Odd-numbered and branched chains decrease melting points.

Crystallization of milkfat largely determines the physical stability of the fat globule and the consistency of high-fat dairy products, but crystal behaviour is also complicated by the wide range of different triglycerides. There are four forms that milkfat crystals can occur in; alpha, ß, ß1, and ß 2, however, the alpha form is the least stable and is rarely observed in slowly cooled fat.

MILK ENZYMES

In the field of biotechnology, there are many industrial applications that result in biotech products that we use every day at home. Some of these are food science applications that utilize enzymes to produce or make improvements in the quality of different foods.

In the dairy industry, some enzymes are required for the production of cheeses, yogurt, and other dairy products, while others are used in a more specialized fashion to improve texture or flavor.

Rennet

Milk contains proteins, specifically caseins, that maintain its liquid form. Proteases are enzymes that are added to milk during cheese production, to hydrolyze caseins, specifically kappa casein, which stabilizes micelle formation preventing coagulation. Rennet and rennin are general terms for any enzyme used to coagulate milk. Technically rennet is also the term for the lining of a calf's fourth stomach.

The most common enzyme isolated from rennet is chymosin. Chymosin can also be obtained from several other animals, microbial or vegetable sources, but indigenous microbial chymosin (from fungi or bacteria) is ineffective for making cheddar and other hard cheeses.

Limited supplies of calf rennet have prompted genetic engineering of microbial chymosin by cloning calf prochymosin genes into bacteria. Bioengineered chymosin may be involved in the production of up to 70% of cheese products. While the use of bioengineered enzymes spares the lives of calves, it presents ethics issues for those opposed to eating foods prepared with GEMs.

Lactalbumin and Lactoglobulin

Milk contains a number of different types of proteins, in addition to the caseins. Cow milk also contains whey proteins such as lactalbumin and lactoglobulin. The denaturing of these whey proteins, using proteases, results in a creamier yogurt product. Destruction of whey proteins is also essential for cheese production.

During the production of soft cheeses, whey is separated from the milk after curdling and may be sold as a nutrient supplement for bodybuilding, weight loss, and lowing blood pressure, among other things. There have even been reports of dietary whey for cancer therapies, and having a role in the induction of insulin production for those with Type 2 diabetes.

Proteases are used to produce hydrolyzed whey protein, which is whey protein broken down into shorter polypeptide sequences. Hydrolyzed whey is less likely to cause allergic reactions and is used to prepare supplements for infant formulas and medical uses.

Lactase

Lactase is a glycoside hydrolase enzyme that cuts lactose into its constituent sugars, galactose, and glucose. Without sufficient production of lactase enzyme in the small intestine, humans become lactose intolerant, resulting in discomfort (cramps, gas, and diarrhea) in the digestive tract upon ingestion of milk products.

Lactase is used commercially to prepare lactose-free products, particularly milk, for such individuals. It is also used in the preparation of ice cream, to make a creamier and sweeter tasting product. Lactase is usually prepared from Kluyveromyces sp. of yeast and Aspergillus sp. of fungi.

Catalase

The enzyme Catalase has found limited use in one particular area of cheese production. Hydrogen peroxide is a potent oxidizer and toxic to cells. It is used instead of pasteurization, when making certain cheeses such as Swiss, in order to preserve natural milk enzymes that are beneficial to the end product and flavor development of the cheese.

These enzymes would be destroyed by the high heat of pasteurization. However, residues of hydrogen peroxide in the milk will inhibit the bacterial cultures that are required for the actual cheese production, so all traces of it must be removed. Catalase enzymes are typically obtained from bovine livers or microbial sources and are added to convert the hydrogen peroxide to water and molecular oxygen.

Lipases

Lipases are used to break down milk fats and give characteristic flavors to cheeses. Stronger flavored cheeses, for example, the Italian cheese, Romano, are prepared using lipases. The flavor comes from the free fatty acids produced when milk fats are hydrolyzed. Animal lipases are obtained from kid, calf, and lamb, while microbial lipase is derived by fermentation with the fungal species Mucor meihei.

Although microbial lipases are available for cheese-making, they are less specific in what fats they hydrolyze, while the animal enzymes are more partial to short and medium-length fats. Hydrolysis of the shorter fats is preferred because it results in the desirable taste of many kinds of cheese. Hydrolysis of the longer chain fatty acids can result in either soapiness or no flavor at all.

MILK MICROBIOLOGY

In addition to being a nutritious food for humans, milk provides a favorable environment for the growth of microorganisms. Yeasts, moulds and a broad spectrum of bacteria can grow in milk, particularly at temperatures above 16 °C.

Microbes can enter milk via the cow, air, feedstuffs, milk handling equipment and the milker. Once microorganisms get into the milk their numbers increase rapidly. It is more effective to exclude microorganisms than to try to control microbial growth once they have entered the milk. Milking equipment should be washed thoroughly before and after use rinsing is not enough. Bacterial types commonly associated with milk are given in Table.

Table: Bacterial types commonly associated with milk.

Pseudomonas	Spoilage
Brucella	Pathogenic
Enterobacteriaceae	Pathogenic and spoilage
Staphylococci	
Staphylococcus aureus	Pathogenic
Streptococcus	
S. agalactiae	Pathogenic
S. thermophilus	Acid fermentation
S. lactis	Acid fermentation
S. lactis-diacetyllatic	Flavour production
S. cremoris	Acid fermentation
Leuconostoc lactis	Acid fermentation
Bacillus cereus	Spoilage
Lactobacillus	
L. lactis	Acid production
L. bulgaricus	Acid production
L. acidophilus	Acid production
Propionibacterium	Acid production
Mycobacterium tuberculosis	Pathogenic

Microbial Growth

Microbial growth can be controlled by cooling the milk. Most microorganisms reproduce slowly in colder environments. Cooling milk also slows chemical deterioration.

The temperature of freshly drawn milk is about 38 °C. Bacteria multiply very rapidly in warm milk and milk sours rapidly if held at these temperatures. If the milk is not cooled and is stored in the shade at an average air temperature of 16 °C, the temperature of the milk will only have fallen to 28 °C after 3 hours. Cooling the milk with running water will reduce the temperature to 16 °C after 1 hour. At this temperature bacterial growth will be reduced and enzyme activity retarded. Thus, milk will keep longer if cooled.

Natural souring of milk may be advantageous: for example, in smallholder butter-making, the acid developed assists in the extraction of fat during churning. The low pH retards growth of lipolytic and proteolytic bacteria and therefore protects the fat and protein in the milk. The acidity of the milk also inhibits the growth of pathogens. It does not, however, retard the growth of molds.

Naturally soured milk is used to make many products, e.g. yoghurt, sour cream, ripened buttermilk and cheese. These products provide ways of preserving milk and are also pleasant to consume. They are produced by the action of fermentative bacteria on lactose and are more readily digested than fresh milk.

The initial microflora of raw milk reflects directly microbial contamination during production. The microflora in milk when it leaves the farm is determined by the temperature to which it has been cooled and the temperature at which it has been stored.

The initial bacterial count of milk may range from less than 1000 cells/ml to 106/ml. High counts (more than 105/ml) are evidence of poor production hygiene. Rapid tests are available for estimating the bacterial quality of milk.

Milk Pasteurisation

Pasteurisation is the process used to destroy bacteria in milk. In pasteurisation, the milk is heated to a temperature sufficient to kill pathogenic bacteria, but well below its boiling point. This also kills many non-pathogenic organisms and thereby extends the storage stability of the milk.

Numerous time/temperature combinations are recommended but the most usual is 72 °C for 15 seconds followed by rapid cooling to below 10 °C. This is normally referred to as High Temperature Short Time (HTST) treatment. It is carried out as a continuous process using a plate heat-exchanger to heat the milk and a holding section to ensure that the milk is completely pasteurised. Milk is normally pasteurised prior to sale as liquid milk. Pasteurisation is used to reduce the microbial counts in milk for cheesemaking, and cream is pasteurised prior to tempering for buttermaking in some factories.

Batch pasteurisation is used where milk quantities are too small to justify the use of a plate heat-exchanger. In batch pasteurisation, fixed quantities of milk are heated to 63 °C and held at this temperature for 30 minutes. The milk is then cooled to 5 °C and packed.

The lower temperature used for batch pasteurisation means that a longer time is required to complete the process 30 minutes at 63 °C, compared with 15 seconds a 72 °C.

Effects of Pasteurisation on Milk

Pasteurisation reduces the cream layer, since some of the fat globule membrane constituents are denatured. This inhibits clustering of the fat globules and consequently reduces the extent of creaming. However, pasteurisation does not reduce the fat content of milk.

Pasteurisation has little effect on the nutritive value of milk. The major nutrients are not altered. There is some loss of vitamin C and B group vitamins, but this is insignificant.

The process kills many fermentative organisms as well as pathogens. Microorganisms that survive pasteurisation are putrefactive. Although pasteurised milk has a storage stability of 2 to 3 days, subsequent deterioration is cause by putrefactive organisms. Thus, pasteurised milk will putrefy rather than develop acidity.

In rural milk processing, many processes depend on the development of acidity, and hence pasteurisation may not be appropriate.

Milk Sterilisation

In pasteurisation, milk receives mild heat treatment to reduce the number of bacteria present. In sterilisation, milk is subjected to severe heat treatment that ensures almost complete destruction of the microbial population. The product is then said to be commercially sterile. Time/temperature treatments of above 100 °C for 15 to 40 minutes are used. The product has a longer shelf life than pasteurised milk.

Another method of sterilisation is ultra-heat treatment, or UHT. In this system, milk is heated under pressure to about 140 °C for 4 seconds. The product is virtually sterile. However, it retains more of the properties of fresh milk than conventionally sterilised milk.

Microbiology of Butter

Butter is made as a means of extracting and preserving milk fat. It can be made directly from milk or by separation of milk and subsequent churning of the cream. In addition to bacteria present in the milk other sources of bacteria in butter are: a) equipment, b) wash water, c) air contamination, d) packing materials, and e) personnel.

Equipment

In smallholder buttermaking, bacterial contamination can come from unclean surfaces, the butter maker and wash water. Packaging materials, cups and leaves are also sources of contaminants. Washing and smoking the churn reduces bacterial numbers. But traditional equipment is often porous and is therefore a reservoir for many organisms.

When butter is made on a larger processing scale, bacterial contamination can come from holding-tank surfaces, the churn and butter-handling equipment.

A wooden churn can be a source of serious bacterial, yeast and mould contamination since these organisms can penetrate the wood, where they can be destroyed only by extreme heat. If a wooden churn has loose bands, cream can enter the crevices between the staves, where it provides a growth medium for bacteria which contaminate subsequent batches of butter. However, if care is taken in cleaning a wooden churn this source of contamination can be controlled. Similar care is required with scotch hands and butterworking equipment.

Wash Water

Wash water can be a source of contamination with both coliform bacteria and bacteria associated with defects in butter. Polluted water supplies can also be a source of pathogens.

Air

Contamination from the air can introduce spoilage organisms: mould spores, bacteria and yeasts can fall on the butter if it is left exposed to the air. Moulds grow rapidly on butter exposed to air.

Packaging

Care is required in the storage and preparation of packaging material. Careless handling of packaging material can be a source of mould contamination.

Personnel

A high standard of personal hygiene is required from people engaged in buttermaking. Personnel pass organisms to butter via the hands, mouth, nasal passage and clothing. Suitable arrangements for disinfecting hands should be provided, and clean working garments should not have contact with other clothes.

Control of Microorganisms in Butter

Salting effectively controls bacterial growth in butter. The salt must be evenly dispersed and worked in well. Salt concentration of 2% adequately dispersed in butter of 16% moisture will result in a 12.5% salt solution throughout the water-in-oil emulsion.

Washing butter does little to reduce microbiological counts. It may be desirable not to wash butter, since washing reduces yield. The acid pH of serum in butter made from ripened cream or sour milk may control the growth of acid-sensitive organisms.

Microbiological analysis of butter usually includes some of the following tests: total bacterial count, yeasts and moulds, coliform estimation and estimation of lipolytic bacteria.

Yeast, mould and coliform estimations are useful for evaluating sanitary practices. The presence of defect producing types can be indicated by estimating the presence of lipolytic organisms.

All butter contains some micro-organisms. However, proper control at every stage of the process can minimise the harmful effects of these organisms.

Standardisation of Milk and Cream

An adjustment of the fat content of cream is required, or if the fat content of whole milk must be reduced to a given level, skim milk must be added. This process is known as standardisation.

Microbial Tests for Raw and Pasteurised Milk

Tests are available to know the milk microbiological quality. Bacteria, coliform and somatic cell counts are frequently used.

Bacteria Count

The total bacteria count is the number of bacteria in a sample that can grow and form countable colonies on Standard Methods Agar after being held at 32 °C for 48 hours.

Coliform Count

The coliform count is the number of colonies in a sample that grow and form distinctive countable colonies on Violet Red Bile Agar after being held at 32 °C for 24 hours. Coliforms are generally only present in food that has been fecally or environmentally contaminated.

Somatic Cell Count

Somatic cells are blood cells that fight infection and occur naturally in milk. The presence of mastitis (an infection of the mammary gland) in the cow will increase the somatic cell count. The somatic cell count can be determined by direct microscopic examination or by electronic instruments designed to count somatic cells.

Antibiotics in Milk

Antibiotics are used to treat mastitis infections. Cows under antibiotic treatment for mastitis infections may have antibiotic residues in their milk, therefore, milk from treated cows is either discarded or collected into a separate tank. Milk containing antibiotic residues is not used for human consumption. The legal standard, as defined by the Food and Drug Administration (FDA), requires that milk contain no detectable antibiotics when analysed using approved test methods. Regulatory action is taken against the farm with the positive antibiotic test.

ULTRA FILTRATION OF MILK

Membrane filtration is a separation process which separates a liquid into two streams by means of a semi-permeable membrane. The two streams are referred to as retentate and permeate. By using membranes with different pore sizes, it is possible to separate specific components of milk and whey. Depending on the application in question, the specified components are either concentrated or removed/reduced.

- Commercial membranes of UF have nominal weight cut-offs at 20,000 – 25,000.

- The protein and fat are retained in the retentate. Lactose, minerals and vitamins are fractioned between the retentate and the permeate.

- Minerals such as Ca, Mg, P and citrate are partly bounded to protein in milk and partly in solution. During UF, the former portion is retained and concentrated and the latter part passes to the retentate.

- Rejection coefficient of the components in whole milk:

  ○ Total solids – 54%

  ○ Protein – 93 %

  ○ Fat – 100 %

  ○ Lactose – 0 %

  ○ Ash – 29 %

  ○ NPN – 62 %

- If during UF, if flux reduces to almost zero because of increase in protein concentration, diafiltration is followed.

- UF milk is useful in formulating reduced or no-lactose dairy products.

Milk Protein Standardization and Fractionation

- UF milk fractionated components can be used to stand the nutritional value of consumer milk or to prepare standardized milk powders, overcoming natural variations in milk composition.

- Increasing the protein content by ultrafiltration makes the milk whiter, and more viscous, the sensory quality more similar to that of higher fat milks.

- An MF/UF process can be used to fractionate non-fat milk into value-added protein ingredients. Resulting ingredients include native casein concentrates (from the retentate), pure milk serum proteins (from the permeate) and individual milk protein isolates that have application as emulsifiers, fortifying proteins and gelling agents.

Fermented Dairy Products

- UF can be used to standardize protein and total solids in milk for use in fermented dairy foods such as cream cheese, yogurt and cottage cheese.

- Fermented products made with UF milk have superior quality and sensory characteristics compared to products made from milk concentrated by conventional methods.

- Membrane filtration helps control quality attributes such as consistency, post-processing acidification and extent of syneresis.

- However, using membrane-processed milk often requires an adjustment in starter culture selection and fermentation conditions.

Cheesemaking

- The cheese industry uses membrane concentrated milk to elevate the solids level of cheesemilk.

- Future applications for membrane processing may include the manufacture of fresh, soft, hard and semi-hard cheese varieties.

- UF concentrated milk, with its fat and protein content concentrated to 3.5X, and a portion of the lactose, ash and water removed, possesses the ideal composition for the potential manufacture of fresh cheeses like ricotta or brine cheeses like Feta.

- Replacing 10-15% of the cheese milk volume with UF milk creates the opportunity to boost total solids, therefore increasing cheese throughput in factory by as much as 18%— subsequently reducing production costs.

- Using concentrated milk could also reduce rennet and starter culture requirements, depending on the application.

- In addition, using UF milk could reduce a cheese plant's wastewater processing costs.

Cream

- Scurlock in 1986 used UF as the 1st stage in the preparation of protein enriched whipping creams. Whole milk concentrated 2 to 5 fold by UF was separated conventionally to produce protein enriched creamswith fat 25 − 45 % (w/w).

Skim Milk Retentate Powder

- Ultra filtered skim milk retentate is spray dried to obtain S.M. retentate powder with 50 − 65 % protein. It shows lower bacterial counts and acidity and an excellent flavour.

- A standardization of the protein levels with WMP with 32 − 35 % protein and skim milk powders with 38 − 41 % protein can be done by using UF skim milk and whey retentates and permeates.

Low Lactose Milk Powder

- It is developed by using low lactose milk powder, lactose content of which can be reduced by 86 %.

- Lactose was replaced by malto-dextrin.

- Na and K salts are added to compensate the loss of milk salts.

Skim Milk based Concentrates

- In-container sterilized milk concentrates are prepared from UF skim milk.

- Such concentrates with edible carbohydrates have TS of 40 % and shelf life over 1 year.

Sweetened Condensed Milk

- UF concentrates are used in making of SCM. Such product has reduced sandiness in the texture.

Ice Cream

- UF can be used to prepare ice cream by using UF skim milk or UF reconstituted skim milk. Ice cream doesn't shows any change in viscosity, specific gravity and overrun but gives improved creamy body and texture.

Non Fat Dairy Coffee Whiteners

- Retentates obtained by UF of skim milk are freeze – dried. The product is comparable to commercial non – dairy coffee creamer and has acceptable dispersibility.

Whey Protein Quarg

- Concentrate obtained by UF of mixture of 80 % whey and 20 % skim milk is heated to obtain firm coagulum. It is also suitable for sweet desserts and puddings.

EFFECT OF MILK CONSTITUENTS ON OPERATION OF MEMBRANE PROCESS

The dairy industry has used membrane processing to clarify, concentrate and fractionate a variety of dairy products. Applying membrane technology to whey processing allowed the production of refined proteins and commercial usage and thus transformed a waste byproduct from cheese production into a valuable product. In addition to whey processing, membrane technology is also used for fluid milk processing with clear advantages. Further, specific milk components can be obtained without causing a phase change to the fluid milk by the addition of heat as in evaporation.

Membrane filtration technologies, such as ultrafiltration and reverse osmosis, are capable of the molecular fractionation of fluids. Milk is ideally suited for processing by membrane filtration because it is a fluid consisting largely of water, lactose, butterfat, and protein molecules. Separation at the molecular level means that butterfat, lactose, and protein can be isolated from one other.

Component in Milk	Average Dimension (nm)
Water	0.2
Lactose	0.5
Casein Proteins	2.0-4.0
Fat	1,000-10,000

Reverse Osmosis

RO appears to be a promising method for concentrating whey with significant savings in the total energy and overall cost. It has been suggested that RO can be used as a pre-concentration step for UF permeate to 20 percent, which would reduce the cost of transportation. The UF permeate containing approximately 4 percent lactose and 1 per cent minerals can be concentrated to 18 percent total solids by RO. From economic point of view, 2 fold concentration (i.e. 50 percent volume reduction) of paneer whey and 2.5 fold concentration (i.e. 60 percent volume reduction) of cow and buffalo cheese Whey.

Ultra Filtration (UF)

Fractionation of whey into protein rich and lactose containing streams is one of the most success-ful industrial applications of UF. Protein content of raw whey can be increased from an initial value of 0.6 per cent to over 20 per cent in the UF step. When whey is concentrated about 20 times by UF, a dry matter content of 18-20 per cent is attainable. It is suggested when UF of whey be carried out for deproteinization for lactose manufacture. Whey protein concentrates (WPC) are powders made by drying of retentate from ultrafiltration of whey. They are described in terms of protein content, percent protein in dry matter, ranging from 35 to 85 percent. To make 35% protein product the liquid whey is concentrated to about 6-fold to an approximate total dry solids content of 9%.

Milk Protein Concentrate using Ultrafiltration

*   Concentrating both casein and whey proteins.

*   Ratio similar to milk.

Table: MPC Composition.

Components (% t/wt)	NFDM	MPC-56	MPC -75	MPC-80
Protein	35	56	75	80
Water	4	5	5	5
Fat	1	1.2	1.5	1.7
Lactose	51.3	31.7	10.9	5.5
Minerals	7.7	8.0	7.2	7.4

Microfiltration

MF can beused to remove large particles: casein fines, micro-organisms or microbial spores, fat globules, somatic cells, phospholipoproteins, particles, etc. from whey. MF separation pro-cess uses porous membranes with a cut off pore sizes in the Region of micron(10-6m)allowing passage of whey proteins but retaining fat globules, microorganisms and somatic cells. Whey usually contains small quantities of fat (in the form of small globules of 0.2 to 1micron) and ca-sein (as fine particulates of 5 to 100micron).Centrifugal separation of whey does not complete-ly remove the fat and casein fines. Thus, when the whey is ultrafiltered, these components can prevent the attainment of high purity, as well as having detrimental effects on the functional properties of WPC. MF can effectively remove substantial quantities of these undesirable com-ponents. Fat: protein ratios of 0.07 to 0.25 in whey can be reduced to 0.001 to 0.003 by MF. In addition some of the precipitated salts may be removed, and there is a considerable reduction in the microbial load. It is reported that 30 to 80 percent residual lipids can be removed from cheddar cheese whey using MF. There is a 1.8 fold increase in the rate of UF of whey proteins when the lipids are removed by MF. When MF is performed on sweet whey as an intermediate step within the UF process, a fat content below 0.4 per cent in 85 percent WPC powder can be achieved.

Microfiltration of Buttermilk

Factors Affecting Membrane Performance Fouling

- Concentration Polarisation
  - Differential solute conc between membrane surface and bulk stream.
  - Reversibly affected by operation parameters.

- Fouling
  - Formation of deposits.
  - Irreversibly affected by operation parameters.

- Membrane Fouling: Two Types
  - Surface (temporary) fouling.
  - Pore (permanent) foul.

- Implications
  - Higher energy consumption.
  - Frequent need for cleaning.
  - Affects membrane durability.
  - Effect on properties and quality of concentrate.
  - Overall economy of the membrane process.

- Surface (Temporary) Fouling
  - Foulant appears an evenly deposited layer on the membrane surface.
  - Can be easily removed by cleaning solution.
  - Permeation rate of membrane can be regenerated by cleaning.
  - Most common type of fouling in UF plant.

- Pore (Permanent) Fouling
  - Particulate matter diffuses into the membrane.
  - Could be caused by the poor quality of the cleaning water.
  - Uneven distribution of the foulant and compression of the separation zone.
  - Flux cannot be regenerated by cleaning.
  - Determines the lifetime of the membrane.

References

- Beecher, Cookson (12 January 2016). "Raw milk's 'explosive growth' comes with costs to the state". Food Safety News. Seattle. Retrieved 24 January 2016

- Bellwood, Peter (2005). "Early Agriculture in the Americas". First Farmers: the origins of agricultural societies. Malden, MA: Blackwell Publushing. Pp. 146–179. ISBN 978-0-631-20566-1

- Dangour AD, Dodhia SK, Hayter A, Allen E, Lock K, Uauy R (September 2009). "Nutritional quality of organic foods: a systematic review". Am. J. Clin. Nutr. 90 (3): 680–5. Doi:10.3945/ajcn.2009.28041. PMID 19640946

- Pearlman, Ann; Bayer, Mary Beth (2010). The Christmas Cookie Cookbook: All the Rules and Delicious Recipes to Start Your Own Holiday Cookie Club. Simon and Schuster. Pp. 197–. ISBN 978-1-4391-7693-1

- Enzymes-used-in-the-dairy-industry: thebalance.com, Retrieved 13 May, 2019

- Dudlicek, James (March 2008). "Renewed focus: a decade after its formation, DFA adjusts its outlook to secure the future for its member owners". Dairy Foods. Retrieved 2008-06-26

- Vitaminsminerals, Milk Composition: milkfacts.info, Retrieved 8 January, 2019

Dairy Products

Some of the commonly used dairy products include yogurt, powdered milk, cream, butter, custard, ice cream, cheese, fermented milk, clabber, soured milk, kefir, kumis, viili and buttermilk. The topics elaborated in this chapter will help in gaining a better perspective about these dairy products.

Dairy product is any of the foods made from milk, including butter, cheese, ice cream, yogurt, and condensed and dried milk.

Milk has been used by humans since the beginning of recorded time to provide both fresh and storable nutritious foods. In some countries almost half the milk produced is consumed as fresh pasteurized whole, low-fat, or skim milk. However, most milk is manufactured into more stable dairy products of worldwide commerce, such as butter, cheese, dried milks, ice cream, and condensed milk.

Cow milk (bovine species) is by far the principal type used throughout the world. Other animals utilized for their milk production include buffalo (China, Egypt, and the Philippines), goats (in the Mediterranean countries), reindeer (in northern Europe), and sheep (in southern Europe).

In the early 1800s the average dairy cow produced less than 1,500 litres of milk annually. With advances in animal nutrition and selective breeding, one cow now produces an average of 6,500 litres of milk a year, with some cows producing up to 10,000 litres. The Holstein-Friesian cow produces the greatest volume, but other breeds such as Ayrshire, Brown Swiss, Guernsey, and Jersey, while producing less milk, are known for supplying milk that contains higher fat, protein, and total solids.

Fat

The fat in milk is secreted by specialized cells in the mammary glands of mammals. It is released as tiny fat globules or droplets, which are stabilized by a phospholipid and protein coat derived from the plasma membrane of the secreting cell. Milk fat is composed mainly of triglycerides—three fatty acid chains attached to a single molecule of glycerol. It contains 65 percent saturated, 32 percent monounsaturated, and 3 percent polyunsaturated fatty acids. The fat droplets carry most of the cholesterol and vitamin A. Therefore, skim milk, which has more than 99.5 percent of the milk fat removed, is significantly lower in cholesterol than whole milk (2 milligrams per 100 grams of milk, compared with 14 milligrams for whole milk) and must be fortified with vitamin A.

Protein

Milk contains a number of different types of proteins, depending on what is required for sustaining the young of the particular species. These proteins increase the nutritional value of milk and other dairy products and provide certain characteristics utilized for many of the processing methods. A major milk protein is casein, which actually exists as a multisubunit protein complex dispersed throughout the fluid phase of milk. Under certain conditions the casein complexes are disrupted, causing curdling of the milk. Curdling results in the separation of milk proteins into two distinct phases, a solid phase (the curds) and a liquid phase (the whey).

Lactose

Lactose is the principal carbohydrate found in milk. It is a disaccharide composed of one molecule each of the monosaccharides (simple sugars) glucose and galactose. Lactose is an important food source for several types of fermenting bacteria. The bacteria convert the lactose into lactic acid, and this process is the basis for several types of dairy products.

In the diet lactose is broken down into its component glucose and galactose subunits by the enzyme lactase. The glucose and galactose can then be absorbed from the digestive tract for use by the body. Individuals deficient in lactase cannot metabolize lactose, a condition called lactose intolerance. The unmetabolized lactose cannot be absorbed from the digestive tract and therefore builds up, leading to intestinal distress.

Vitamins and Minerals

Milk is a good source of many vitamins. However, its vitamin C (ascorbic acid) content is easily destroyed by heating during pasteurization. Vitamin D is formed naturally in milk fat by ultraviolet irradiation but not in sufficient quantities to meet human nutritional needs. Beverage milk is commonly fortified with the fat-soluble vitamins A and D. In the United States the fortification of skim milk and low-fat milk with vitamin A (in water-soluble emulsified preparations) is required by law.

Milk also provides many of the B vitamins. It is an excellent source of riboflavin (B_2) and provides lesser amounts of thiamine (B_1) and niacin. Other B vitamins found in trace amounts are pantothenic acid, folic acid, biotin, pyridoxine (B_6), and vitamin B_{12}.

Milk is also rich in minerals and is an excellent source of calcium and phosphorus. It also contains trace amounts of potassium, chloride, sodium, magnesium, sulfur, copper, iodine, and iron. A lack of adequate iron is said to keep milk from being a complete food.

Physical and Biochemical Properties

Milk contains many natural enzymes, and other enzymes are produced in milk as a result of bacterial growth. Enzymes are biological catalysts capable of producing chemical changes in organic substances. Enzyme action in milk systems is extremely important for its effect on the flavour and body of different milk products. Lipases (fat-splitting enzymes), oxidases, proteases (protein-splitting enzymes), and amylases (starch-splitting enzymes) are among the more important enzymes that occur naturally in milk. These classes of enzymes are also produced in milk by microbiological action. In addition, the proteolytic enzyme (i.e., protease) rennin, produced in calves stomachs to coagulate milk protein and aid in nutrient absorption, is used to coagulate milk for manufacturing cheese.

The coagulation of milk is an irreversible change of its protein from a soluble or dispersed state to an agglomerated or precipitated condition. Its appearance may be associated with spoilage, but coagulation is a necessary step in many processing procedures. Milk may be coagulated by rennin or other enzymes, usually in conjunction with heat. Left unrefrigerated, milk may naturally sour or co-agulate by the action of lactic acid, which is produced by lactose-fermenting bacteria. This principle is utilized in the manufacture of cottage cheese. When milk is pasteurized and continuously refrigerated for two or three weeks, it may eventually coagulate or spoil owing to the action of psychrophilic or proteolytic organisms that are normally present or result from postpasteurization contamination.

Milk fat is present in milk as an emulsion in a water phase. Finely dispersed fat globules in this emulsion are stabilized by a milk protein membrane, which permits the fat to clump and rise. The rising action is called creaming and is expected in all unhomogenized milk. In the United States, when paper cartons supplanted glass bottles, consumers stopped the practice of skimming cream from the top. Processors then introduced homogenization, a method of preventing gravity separation by forcing milk through very small openings under pressure, thus reducing fat globules to one-tenth their original size. Homogenization is practiced in many dairy processes in order to improve the physical properties of products.

Milk and other dairy products are very susceptible to developing off-flavours. Some flavours, given such names as "feed," "barny," or "unclean," are absorbed from the food ingested by the cow and from the odours in its surroundings. Others develop through microbial action due to growth of bacteria in large numbers. Chemical changes can also take place through enzyme action, contact with metals (such as copper), or exposure to sunlight or strong fluorescent light. Quality-control directors are constantly striving to avoid off-flavours in milk and other dairy foods.

Fresh Fluid Milk

Fresh fluid milk requires the highest-quality raw milk and is generally designated as Grade "A." This grade requires a higher level of sanitation and inspection on the farm than is necessary for "manufacturing grade" milk.

Quality Concerns

Raw milk is a potentially dangerous food that must be processed and protected to assure its safety for humans. While most bovine diseases, such as brucellosis and tuberculosis, have been eliminated, many potential human pathogens inhabit the dairy farm environment. Therefore, it is essential that all milk be either pasteurized or (in the case of cheese) held for at least 60 days if made from raw milk. While milk from healthy cows is often totally bacteria-free, that condition quickly changes when milk is exposed to the farm environment.

Milk received at the processing plant is tested before being unloaded from either farm-based tank trucks or over-the-road tankers. The milk is checked for odour, appearance, proper temperature, acidity, bacteria, and the presence of drug residues. These tests take no longer than 10 to 15 minutes. If the tank load passes these tests, the milk is pumped into the plant's refrigerated storage tanks. The milk is then stored for the shortest possible time.

Pasteurization

Pasteurization is most important in all dairy processing. It is the biological safeguard which ensures that all potential pathogens are destroyed. Extensive studies have determined that heating milk to 63 °C (145 °F) for 30 minutes or 72 °C (161 °F) for 15 seconds kills the most resistant harmful bacteria. In actual practice these temperatures and times are exceeded, thereby not only ensuring safety but also extending shelf life.

Most milk today is pasteurized by the continuous high-temperature short-time (HTST) method (72 °C or 161 °F for 15 seconds or above). The HTST method is conducted in a series of stainless steel plates and tubes, with the hot pasteurized milk on one side of the plate being cooled by the incoming raw milk on the other side. This "regeneration" can be more than 90 percent efficient and greatly reduces the cost of heating and cooling. There are many fail-safe controls on an approved pasteurizer system to ensure that all milk is completely heated for the full time and temperature requirement. If the monitoring instruments detect that something is wrong, an automatic flow diversion valve will prevent the milk from moving on to the next processing stage. Higher temperatures and sometimes longer holding times are required for the pasteurization of milk or cream with a high fat or sugar content.

Pasteurized milk is not sterile and is expected to contain small numbers of harmless bacteria. Therefore, the milk must be immediately cooled to below 4.4 °C (40 °F) and protected from any outside contamination. The shelf life for high-quality pasteurized milk is about 14 days when properly refrigerated.

Extended shelf life can be achieved through ultrapasteurization. In this case, milk is heated to 138 °C (280 °F) for two seconds and aseptically placed in sterile conventional milk containers. Ultrapasteurized milk and cream must be refrigerated and will last at least 45 days. This process does minimal damage to the flavour and extends the shelf life of slow-selling products such as cream, eggnog, and lactose-reduced milks.

Ultrahigh-temperature (UHT) pasteurization is the same heating process as ultrapasteurization (138 °C or 280 °F for two seconds), but the milk then goes into a more substantial container—either a sterile five-layer laminated "box" or a metal can. This milk can be stored without refrigeration and has a shelf life of six months to a year. Products handled in this manner do not taste as fresh, but they are useful as an emergency supply or when refrigeration is not available.

Separation

Most modern plants use a separator to control the fat content of various products. A separator is a high-speed centrifuge that acts on the principle that cream or butterfat is lighter than other components in milk. The specific gravity of skim milk is 1.0358, specific gravity of heavy cream 1.0083. The heart of the separator is an airtight bowl with funnellike stainless steel disks. The bowl is spun at a high speed (about 6,000 revolutions per minute), producing centrifugal forces of 4,000 to

5,000 times the force of gravity. Centrifugation causes the skim, which is denser than cream, to collect at the outer wall of the bowl. The lighter part (cream) is forced to the centre and piped off for appropriate use.

An additional benefit of the separator is that it also acts as a clarifier. Particles even heavier than the skim, such as sediment, somatic cells, and some bacteria, are thrown to the outside and collected in pockets on the side of the separator. This material, known as "separator sludge," is discharged periodically and sometimes automatically when buildup is sensed.

Most separators are controlled by computers and can produce milk of almost any fat content. Current standards generally set whole milk at 3.25 percent fat, low-fat at 1 or 2 percent, and skim at less than 0.5 percent. (Most skim milk is actually less than 0.01 percent fat).

Homogenization

Milk is homogenized to prevent fat globules from floating to the top and forming a cream layer or cream plug. Homogenizers are simply heavy-duty, high-pressure pumps equipped with a special valve at the discharge end. They are designed to break up fat globules from their normal size of up to 18 micrometres to less than 2 micrometres in diameter (a micrometre is one-millionth of a metre). Hot milk (with the fat in liquid state) is pumped through the valve under high pressure, resulting in a uniform and stable distribution of fat throughout the milk.

Two-stage homogenization is sometimes practiced, during which the milk is forced through a second homogenizer valve or a breaker ring. The purpose is to break up fat clusters or clumps and thus produce a more uniform product with a slightly reduced viscosity.

Homogenization is considered successful when there is no visible separation of cream and the fat content in the top 100 millilitres of milk in a one-litre bottle does not differ by more than 10 percent from the bottom portion after standing 48 hours.

In addition to avoiding a cream layer, other benefits of homogenized milk include a whiter appearance, richer flavour, more uniform viscosity, better "whitening" in coffee, and softer curd tension (making the milk more digestible for humans). Homogenization is also essential for providing improved body and texture in ice cream, as well as numerous other products such as half-and-half, cream cheese, and evaporated milk.

Packaging

Until the mid 1880s milk was dipped from large cans into the consumer's own containers. The glass milk bottle was invented in 1884 and became the main container of retail distribution until World War II, when wax-coated paper containers were introduced. Plastic-coated paper followed and became the predominate container. Today more than 75 percent of retail sales are in translucent plastic jugs. Glass bottles make up less than 0.5 percent of the business and are used mostly at dairy stores and for home delivery.

Modern packaging machines are self-cleaning and provide an aseptic environment for milk packaging. Their improved design has allowed milk to remain fresh for at least 14 days and has made it possible for use with ultrapasteurizing equipment for extended shelf-life applications.

Specialty Milks

Many specialty milks are now available (even in remote areas) as a result of the 45-day refrigerated shelf life of ultrapasteurized milk. One of the most useful products, lactose-reduced milk, is available in both nonfat and low-fat composition as well as in many flavoured versions. The lactose (milk sugar) is reduced by 70 to 100 percent, making it possible for lactose-intolerant individuals to enjoy the benefits of milk in their diets. Lactose reduction is accomplished by subjecting the appropriate milk to the action of the enzyme lactase in a refrigerated tank for approximately 24 hours. The enzyme breaks down the lactose to more readily digestible glucose and galactose. The reaction is halted when the lactose is consumed or when the milk is heat-treated. The resulting beverage is sweeter than regular milk but acceptable for most uses.

Other specialty milks include calcium-fortified, special and seasonal flavours (e.g., eggnog), and high-volume flavoured milk shakes (frequently served in schools).

Condensed and Dried Milk

Condensed and Evaporated Milk

Whole, low-fat, and skim milks, as well as whey and other dairy liquids, can be efficiently concentrated by the removal of water, using heat under vacuum. Since reducing atmospheric pressure lowers the temperature at which liquids boil, the water in milk is evaporated without imparting a cooked flavour. Water can also be removed by ultrafiltration and reverse osmosis, but this membrane technology is more expensive. Usually about 60 percent of the water is removed, which reduces storage space and shipping costs. Whole milk, when concentrated, usually contains 7.5 percent milk fat and 25.5 percent total milk solids. Skim milk can be condensed to approximately 20 to 40 percent solids, depending on the buyer's needs.

Condensed milk is often sold in refrigerated tank-truck loads to manufacturers of candy, bakery goods, ice cream, cheese, and other foods. When preserved by heat in individual cans, it is usually called "evaporated milk." In this process the concentrated milk is homogenized, fortified with vitamin D (A and D in evaporated skim milk), and sealed in a can sized for the consumer. A stabilizer, such as disodium phosphate or carrageenan, is also added to keep the product from separating during processing and storage. The sealed can is then sterilized at 118 °C (244 °F) for 15 minutes, cooled, and labeled. Evaporated milk keeps indefinitely, although staling and browning may occur after a year.

New ultrahigh-temperature (UHT) processing and aseptic filling of foil-lined cardboard or metal cans is also practiced. Although this process is more costly, the scorched flavour is not as pronounced as with conventionally processed evaporated milk.

Sweetened condensed milk is also made by partially removing the water (as in evaporated milk) and adding sugar. The final product contains about 8.5 percent milk fat and at least 28 percent total milk solids. Sugar is added in sufficient amount to prevent bacterial action and subsequent spoilage. Usually, at least 60 percent sugar in the water phase is required to provide sufficient osmotic pressure for prevention of bacterial growth. Because sweetened condensed milk (or skim milk) is preserved by sugar, the milk merely needs to be pasteurized before being placed in a sanitary container (usually a metal can).

Dry Milk Products

Milk and by-products of milk production are often dried to reduce weight, to aid in shipping, to extend shelf life, and to provide a more useful form as an ingredient for other foods. In addition to skim and whole milk, a variety of useful dairy products are dried, including buttermilk, malted milk, instant breakfast, sweet cream, sour cream, butter powder, ice cream mix, cheese whey, coffee creamer, dehydrated cheese products, lactose, and caseinates. Many drying plants are built in conjunction with a butter-churning plant. These plants utilize the skim milk generated from the separated cream and the buttermilk produced from churning the butter. Most products are dried to less than 4 percent moisture to prevent bacterial growth and spoilage. However, products containing fat lose their freshness rather quickly owing to the oxidation of fatty acids, leading to rancidity.

Two types of dryers are used in the production of dried milk products—drum dryers and spray dryers. Each dryer has certain advantages.

Drum Dryers

The simplest and least expensive is the drum, or roller, dryer. It consists of two large steel cylinders that turn toward each other and are heated from the inside by steam. The concentrated product is applied to the hot drum in a thin sheet that dries during less than one revolution and is scraped from the drum by a steel blade. The flakelike powder dissolves poorly in water but is often preferred in certain bakery products. Drum dryers are also used to manufacture animal feed where texture, flavour, and solubility are not a major consideration.

Spray Dryers

Spray dryers are more commonly used since they do less heat damage and produce more soluble products. Concentrated liquid dairy product is sprayed in a finely atomized form into a stream of hot air. The air may be heated by steam-heated "radiators" or directly by sulfur-free natural gas. The drying chamber may be rectangular (the size of a living room), conical, or silo-shaped (up to five stories high). The powder passes from the drying chamber through a series of cyclone collectors and is usually placed in plastic-lined, heavy-duty paper bags.

Spray-dried milk is also difficult to reconstitute or mix with water. Therefore, a process called agglomeration was developed to "instantize" the powder, or make it more soluble. This process involves rewetting the fine, spray-dried powder with water to approximately 8 to 15 percent moisture and following up with a second drying cycle. The powder is now granular and dissolves very well in water. Virtually all retail packages of nonfat dry milk powder are instantized in this manner.

YOGURT

Yogurt, also spelled yoghurt, yogourt or yoghourt, is a food produced by bacterial fermentation of milk. The bacteria used to make yogurt are known as yogurt cultures. The fermentation of lactose by these bacteria produces lactic acid, which acts on milk protein to give yogurt its texture and characteristic tart flavor. Cow's milk is commonly available worldwide and, as such, is the milk most commonly used to make yogurt. Milk from water buffalo, goats, ewes, mares, camels, and yaks is also used to produce yogurt where available locally. The milk used may be homogenized or not, even pasteurized or raw. Each type of milk produces substantially different results.

Yogurt is produced using a culture of Lactobacillus delbrueckii subsp. bulgaricus and Streptococcus thermophilus bacteria. In addition, other lactobacilli and bifidobacteria are sometimes added during or after culturing yogurt. Some countries require yogurt to contain a certain amount of colony-forming units (CFU) of bacteria; in China, for example, the requirement for the number of lactobacillus bacteria is at least 1 million CFU per millilitre.

To produce yogurt, milk is first heated, usually to about 85 °C (185 °F), to denature the milk proteins so that they do not form curds. After heating, the milk is allowed to cool to about 45 °C (113 °F). The bacterial culture is mixed in, and that temperature of 45 °C is maintained for 4 to 12 hours to allow fermentation to occur.

Market and Consumption

In 2017, the average American ate 13.7 pounds of yogurt and that figure has been declining since 2014.

Sale of yogurt was down 3.4 percent over the 12 months ending in February 2019. The decline of Greek-style yogurt has allowed Icelandic style yogurt to gain a foothold in the United States with sales of the Icelandic style yogurt increasing 24 percent in 2018 to $173 million.

Nutrition and Health

Yogurt (plain yogurt from whole milk) is 81% water, 9% protein, 5% fat, and 4% carbohydrates, including 4% sugars (table). A 100-gram amount provides 406 kilojoules (97 kcal) of dietary energy. As a proportion of the Daily Value (DV), a serving of yogurt is a rich source of vitamin B_{12} (31% DV) and riboflavin (23% DV), with moderate content of protein, phosphorus, and selenium (14 to 19% DV; table).

Comparison of whole milk and plain yogurt from whole milk, one cup (245 g) each		
Property	Milk	Yogurt
Energy	610 kJ (146 kcal)	620 kJ (149 kcal)
Total carbohydrates	12.8 g	12 g
Total fat	7.9 g	8.5 g
Cholesterol	24 mg	32 mg
Protein	7.9 g	9 g
Calcium	276 mg	296 mg
Phosphorus	222 mg	233 mg
Potassium	349 mg	380 mg
Sodium	98 mg	113 mg
Vitamin A	249 IU	243 IU
Vitamin C	0.0 mg	1.2 mg
Vitamin D	96.5 IU	~

Vitamin E	0.1 mg	0.1 mg
Vitamin K	0.5 µg	0.5 µg
Thiamine	0.1 mg	0.1 mg
Riboflavin	0.3 mg	0.3 mg
Niacin	0.3 mg	0.2 mg
Vitamin B6	0.1 mg	0.1 mg
Folate	12.2 µg	17.2 µg
Vitamin B12	1.1 µg	0.9 µg
Choline	34.9 mg	37.2 mg
Betaine	1.5 mg	~
Water	215 g	215 g
Ash	1.7 g	1.8 g

Tilde (~) represents missing or incomplete data. The above shows little difference exists between whole milk and yogurt made from whole milk with respect to the listed nutritional constituents.

Because it may contain live cultures, yogurt is often associated with probiotics, which have been postulated as having positive effects on immune, cardiovascular or metabolic health. However, to date high-quality clinical evidence has been insufficient to conclude that consuming yogurt lowers the risk of diseases or otherwise improves health.

Varieties and Presentation

Tzatziki or cacık is a meze made with yogurt,
cucumber, olive oil and fresh mint or dill.

Dahi is a yogurt from the Indian subcontinent, known for its characteristic taste and consistency. The word dahi seems to be derived from the Sanskrit word dadhi ("sour milk"), one of the five elixirs, or panchamrita, often used in Hindu ritual. Sweetened dahi (mishti doi or meethi dahi) is common in eastern parts of India, made by fermenting sweetened milk. While cow's milk is considered sacred and is currently the primary ingredient for yogurt, goat and buffalo milk were widely used in the past, and valued for the fat content.

Dadiah or dadih is a traditional West Sumatran yogurt made from water buffalo milk, fermented in bamboo tubes. Yogurt is common in Nepal, where it is served as both an appetizer and dessert. Locally called dahi, it is a part of the Nepali culture, used in local festivals, marriage ceremonies, parties, religious occasions, family gatherings, and so on. One Nepalese yogurt is called juju dhau, originating from the city of Bhaktapur. In Tibet, yak milk (technically dri milk, as the word yak refers to the male animal) is made into yogurt (and butter and cheese) and consumed.

In Northern Iran, Mâst Chekide is a variety of kefir yogurt with a distinct sour taste. It is usually mixed with a pesto-like water and fresh herb purée called delal. Common appetizers are spinach or eggplant borani, Mâst-o-Khiâr with cucumber, spring onions and herbs, and Mâst-Musir with wild shallots. In the summertime, yogurt and ice cubes are mixed together with cucumbers, raisins, salt, pepper and onions and topped with some croutons made of Persian traditional bread and served as a cold soup. Ashe-Mâst is a warm yogurt soup with fresh herbs, spinach and lentils. Even the leftover water extracted when straining yogurt is cooked to make a sour cream sauce called kashk, which is usually used as a topping on soups and stews.

Matsoni is a Georgian yogurt in the Caucasus and Russia. Tarator and Cacık are cold soups made from yogurt during summertime in eastern Europe. They are made with ayran, cucumbers, dill, salt, olive oil, and optionally garlic and ground walnuts. Tzatziki in Greece and milk salad in Bulgaria are thick yogurt-based salads similar to tarator.

Khyar w Laban (cucumber and yogurt salad) is a dish in Lebanon and Syria. Also, a wide variety of local Lebanese and Syrian dishes are cooked with yogurt like "Kibbi bi Laban" Rahmjoghurt, a creamy yogurt with much higher fat content (10%) than many yogurts offered in English-speaking countries. Dovga, a yogurt soup cooked with a variety of herbs and rice, is served warm in winter or refreshingly cold in summer. Jameed, yogurt salted and dried to preserve it, is consumed in Jordan. Zabadi is the type of yogurt made in Egypt, usually from the milk of the Egyptian water buffalo. It is particularly associated with Ramadan fasting, as it is thought to prevent thirst during all-day fasting.

Sweetened and Flavored

To offset its natural sourness, yogurt is also sold sweetened, sweetened and flavored or in containers with fruit or fruit jam on the bottom. The two styles of yogurt commonly found in the grocery store are set-style yogurt and Swiss-style yogurt. Set-style yogurt is poured into individual containers to set, while Swiss-style yogurt is stirred prior to packaging. Either may have fruit added to increase sweetness.

Lassi and moru are common beverages in India. Lassi is stirred liquified yogurt that is either salted or sweetened with sugar commonly, less commonly honey and often combined with fruit pulp to create flavored lassi. Mango lassi is a western favorite, as is coconut lassi. Consistency can vary widely, with urban and commercial lassis being of uniform texture through being processed, whereas rural and rustic lassi has discernible curds in it, and sometimes has cream added or removed. Moru is a South Indian summer drink, meant to keep drinkers hydrated through the hot and humid summers of the South. It is prepared by considerably thinning down yogurt with water, adding salt (for electrolyte balance) and spices, usually green chili peppers, asafoetida, curry leaves and mustard.

Large amounts of sugar or other sweeteners for low-energy yogurts are often used in commercial yogurt. Some yogurts contain added modified starch, pectin (found naturally in fruit), and gelatin to create thickness and creaminess artificially at lower cost. This type of yogurt is also marketed under the name Swiss-style, although it is unrelated to the way yogurt is eaten in Switzerland. Some yogurts, often called "cream line", are made with whole milk which has not been homogenized so the cream rises to the top. In the UK, Ireland, France and United States, sweetened, flavored yogurt is common, typically sold in single-serving plastic cups. Common flavors include vanilla, honey, and toffee, and fruit such as strawberry, cherry, blueberry, blackberry, raspberry, mango and peach. In the early twenty-first century yogurt flavors inspired by desserts, such as chocolate or cheesecake, have been available. There is concern about the health effects of sweetened yogurt, due to its high sugar content.

Straining

A coffee filter used to strain yogurt in a home refrigerator.

Strained yogurt has been strained through a filter, traditionally made of muslin and more recently of paper or non-muslin cloth. This removes the whey, giving a much thicker consistency. Strained yogurt is made at home, especially if using skimmed milk which results in a thinner consistency. Yogurt that has been strained to filter or remove the whey is known as Labneh in Middle Eastern countries. It has a consistency between that of yogurt and cheese. It may be used for sandwiches in Middle Eastern countries. Olive oil, cucumber slices, olives, and various green herbs may be added. It can be thickened further and rolled into balls, preserved in olive oil, and fermented for a few more weeks. It is sometimes used with onions, meat, and nuts as a stuffing for a variety of pies or kibbeh balls.

Some types of strained yogurts are boiled in open vats first, so that the liquid content is reduced. The East Indian dessert, a variation of traditional dahi called mishti dahi, offers a thicker, more custard-like consistency, and is usually sweeter than western yogurts. Strained yogurt is also enjoyed in Greece and is the main component of *tzatziki* (from Turkish "cacık"), a well-known accompaniment to gyros and souvlaki pita sandwiches: it is a yogurt sauce or dip made with the addition of grated cucumber, olive oil, salt and, optionally, mashed garlic. Srikhand, a dessert in India, is made from strained yogurt, saffron, cardamom, nutmeg and sugar and sometimes fruits such as mango or pineapple.

In North America, strained yogurt is commonly called "Greek yogurt" and in Britain as "Greek-style yogurt". In Britain the name "Greek" may only be applied to yogurt made in Greece.

Beverages

Ayran, doogh ("dawghe" in Neo-Aramaic) or dhallë is a yogurt-based, salty drink. It is made by mixing yogurt with water and (sometimes) salt.

Borhani (or burhani) is a spicy yogurt drink from Bangladesh. It is usually served with kacchi biryani at weddings and special feasts. Key ingredients are yogurt blended with mint leaves (mentha), mustard seeds and black rock salt (Kala Namak). Ground roasted cumin, ground white pepper, green chili pepper paste and sugar are often added.

Lassi is a yogurt-based beverage that is usually slightly salty or sweet, and may be commercially flavored with rosewater, mango or other fruit juice. Salty lassi is usually flavored with ground, roasted cumin and red chilies, may be made with buttermilk.

An unsweetened and unsalted yogurt drink usually called simply *jogurt* is consumed with *burek* and other baked goods. Sweetened yogurt drinks are the usual form in Europe (including the UK) and the US, containing fruit and added sweeteners. These are typically called "drinkable yogurt". Also available are "yogurt smoothies", which contain a higher proportion of fruit and are more like smoothies.

Plant Milk Yogurt

A variety of plant milk yogurts appeared in the 2000s, using soy milk, rice milk, and nut milks such as almond milk and coconut milk. These yogurts are suitable for vegans, people with intolerance to dairy milk, and those who prefer plant-based products.

Homemade

Commercially available home yogurt maker.

Yogurt is made by heating milk to a temperature that denaturates its proteins (scalding), essential for making yogurt, cooling it to a temperature that will not kill the live microorganisms that turn the milk into yogurt, inoculating certain bacteria (starter culture), usually Streptococcus thermophilus and Lactobacillus bulgaricus, into the milk, and finally keeping it warm for several hours. The milk may be held at 85 °C (185 °F) for a few minutes, or boiled (giving a somewhat different result). It must be cooled to 50 °C (122 °F) or somewhat less, typically 40–46 °C (104–115 °F). Starter culture must then be mixed in well, and the mixture must be kept undisturbed and warm for some time, anywhere between 5 and 12 hours. Longer fermentation times produces a more acidic yogurt. The starter culture may be a small amount of live (not sterilized) existing yogurt or commercially available dried starter culture.

Milk with a higher concentration of solids than normal milk may be used; the higher solids content produces a firmer yogurt. Solids can be increased by adding dried milk. The yogurt-making process provides two significant barriers to pathogen growth, heat and acidity (low pH). Both are necessary to ensure a safe product. Acidity alone has been questioned by recent outbreaks of food poisoning by E. coli O157:H7 that is acid-tolerant. E. coli O157:H7 is easily destroyed by pasteurization (heating); the initial heating of the milk kills pathogens as well as denaturing proteins. The microorganisms that turn milk into yogurt can tolerate higher temperatures than most pathogens, so that a suitable temperature not only encourages the formation of yogurt, but inhibits pathogenic microorganisms. Once the yogurt has formed it can, if desired, be strained to reduce the whey content and thicken it.

Commercial Yogurt

Two types of yogurt are supported by the Codex Alimentarius for import and export, implemented similarly by the US Food and Drug Administration.

- Pasteurized yogurt ("heat treated fermented milk") is yogurt pasteurized to kill bacteria.

- Probiotic yogurt (labeled as "live yogurt" or "active yogurt") is yogurt pasteurized to kill bacteria, with *Lactobacillus* added in measured units before packaging.

- Yogurt probiotic drink is a drinkable yogurt pasteurized to kill bacteria, with *Lactobacillus* added before packaging.

Research suggests Homemade Yogurt and Live Yogurt are much more beneficial than 'Heat Treated Fermented Milk' (Pasteurized Yogurt). European Food Safety Authority has confirmed the probiotic benefits of Yogurt containing 10^8 CFU 'live Lactobacilli'.

Lactose Intolerance

Lactose intolerance is a condition in which people have symptoms due to the decreased ability to digest lactose, a sugar found in dairy products. In 2010, the European Food Safety Authority (EFSA) determined that lactose intolerance can be alleviated by ingesting live yogurt cultures (lactobacilli) that are able to digest the lactose in other dairy products. The scientific review by EFSA enabled yogurt manufacturers to use a health claim on product labels, provided that the "yogurt should contain at least 108 CFU live starter microorganisms (Lactobacillus delbrueckii subsp. bulgaricus and Streptococcus thermophilus) per gram. The target population is individuals with lactose maldigestion."

Yogurt Production

Yogurt is a fermented milk product that contains the characteristic bacterial cultures Lactobacillus bulgaricus and Streptococcus thermophilus. All yogurt must contain at least 8.25% solids not fat. Full fat yogurt must contain not less than 3.25% milk fat, lowfat yogurt not more than 2% milk fat, and nonfat yogurt less than 0.5% milk.

The two styles of yogurt commonly found in the grocery store are set type yogurt and swiss style yogurt. Set type yogurt is when the yogurt is packaged with the fruit on the bottom of the cup and the yogurt on top. Swiss style yogurt is when the fruit is blended into the yogurt prior to packaging.

Ingredients

The main ingredient in yogurt is milk. The type of milk used depends on the type of yogurt – whole milk for full fat yogurt, lowfat milk for lowfat yogurt, and skim milk for nonfat yogurt. Other dairy ingredients are allowed in yogurt to adjust the composition, such as cream to adjust the fat content, and nonfat dry milk to adjust the solids content. The solids content of yogurt is often adjusted above the 8.25% minimum to provide a better body and texture to the finished yogurt. The CFR contains a list of the permissible dairy ingredients for yogurt.

Stabilizers may also be used in yogurt to improve the body and texture by increasing firmness, preventing separation of the whey (syneresis), and helping to keep the fruit uniformly mixed in the yogurt. Stabilizers used in yogurt are alginates (carageenan), gelatins, gums (locust bean, guar), pectins, and starch. Sweeteners, flavors and fruit preparations are used in yogurt to provide variety to the consumer. A list of permissible sweeteners for yogurt is found in the CFR.

Bacterial Cultures

The main (starter) cultures in yogurt are Lactobacillus bulgaricus and Streptococcus thermophilus. The function of the starter cultures is to ferment lactose (milk sugar) to produce lactic acid. The increase in lactic acid decreases pH and causes the milk to clot, or form the soft gel that is characteristic of yogurt. The fermentation of lactose also produces the flavor compounds that are characteristic of yogurt. Lactobacillus bulgaricus and Streptococcus thermophilus are the only 2 cultures required by law (CFR) to be present in yogurt.

Other bacterial cultures, such as Lactobacillus acidophilus, Lactobacillus subsp. casei, and Bifido-bacteria may be added to yogurt as probiotic cultures. Probiotic cultures benefit human health by improving lactose digestion, gastrointestinal function, and stimulating the immune system.

General Manufacturing Procedure

The following flow chart and discussion provide a general outline of the steps required for making yogurt.

General Yogurt Processing Steps:

- Adjust Milk Composition & Blend Ingredients,
- Pasteurize Milk,
- Homogenize,
- Cool Milk,
- Inoculate with Starter Cultures,
- Hold,

- Cool,

- Add Flavors & Fruit,

- Package.

Adjust Milk Composition and Blend Ingredients

Milk composition may be adjusted to achieve the desired fat and solids content. Often dry milk is added to increase the amount of whey protein to provide a desirable texture. Ingredients such as stabilizers are added at this time.

Pasteurize Milk

The milk mixture is pasteurized at 185 °F (85 °C) for 30 minutes or at 203 °F (95 °C) for 10 minutes. A high heat treatment is used to denature the whey (serum) proteins. This allows the proteins to form a more stable gel, which prevents separation of the water during storage. The high heat treatment also further reduces the number of spoilage organisms in the milk to provide a better environment for the starter cultures to grow. Yogurt is pasteurized before the starter cultures are added to ensure that the cultures remain active in the yogurt after fermentation to act as probiotics; if the yogurt is pasteurized after fermentation the cultures will be inactivated.

Homogenize

The blend is homogenized (2000 to 2500 psi) to mix all ingredients thoroughly and improve yogurt consistency.

Cool Milk

The milk is cooled to 108 °F (42 °C) to bring the yogurt to the ideal growth temperature for the starter culture.

Inoculate with Starter Cultures

The starter cultures are mixed into the cooled milk.

Hold

The milk is held at 108 °F (42 °C) until a pH 4.5 is reached. This allows the fermentation to progress to form a soft gel and the characteristic flavor of yogurt. This process can take several hours.

Cool

The yogurt is cooled to 7 °C to stop the fermentation process.

Add Fruit and Flavors

Fruit and flavors are added at different steps depending on the type of yogurt. For set style yogurt the fruit is added in the bottom of the cup and then the inoculated yogurt is poured on top and the yogurt is fermented in the cup. For swiss style yogurt the fruit is blended with the fermented, cooled yogurt prior to packaging.

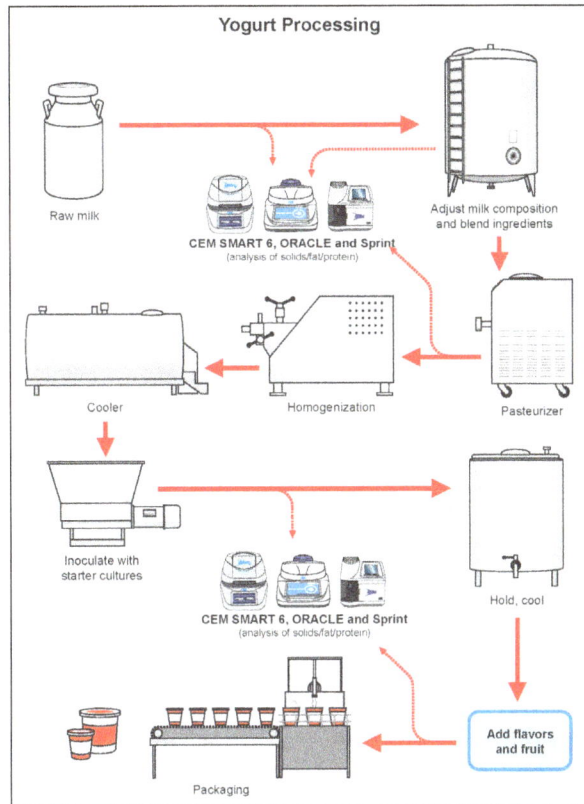

POWDERED MILK

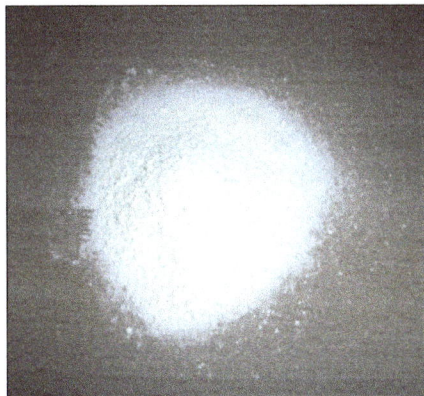

Powdered milk.

Powdered milk or dried milk is a manufactured dairy product made by evaporating milk to dryness. One purpose of drying milk is to preserve it; milk powder has a far longer shelf life than liquid milk and does not need to be refrigerated, due to its low moisture content. Another purpose is to reduce its bulk for economy of transportation. Powdered milk and dairy products include such items as dry whole milk, nonfat (skimmed) dry milk, dry buttermilk, dry whey products and dry dairy blends. Many dairy products exported conform to standards laid out in Codex Alimentarius. Many forms of milk powder are traded on exchanges.

Powdered milk is used for food and health (nutrition), and also in biotechnology (saturating agent).

Food and Health Uses

Powdered milk.

Powdered milk is frequently used in the manufacture of infant formula, confectionery such as chocolate and caramel candy, and in recipes for baked goods where adding liquid milk would render the product too thin. Powdered milk is also widely used in various sweets such as the famous Indian milk balls known as gulab jamun and a popular Indian sweet delicacy (sprinkled with desiccated coconut) known as chum chum (made with skim milk powder). Many no-cook recipes that use nut butters use powdered milk to prevent the nut butter from turning liquid by absorbing the oil.

Powdered milk is also a common item in UN food aid supplies, fallout shelters, warehouses, and wherever fresh milk is not a viable option. It is widely used in many developing countries because of reduced transport and storage costs (reduced bulk and weight, no refrigerated vehicles). Like other dry foods, it is considered nonperishable, and is favored by survivalists, hikers, and others requiring nonperishable, easy-to-prepare food.

Because of its resemblance to cocaine and other drugs, powdered milk is sometimes used in filmmaking as a non-toxic prop that may be insufflated.

Reconstitution

The weight of nonfat dry milk (NFDM) to use is about 10% of the water weight. Alternatively, when measuring by volume rather than weight, one cup of fluid milk from powdered milk requires one cup of water and one-third cup of powdered milk.

Nutritional Value

Milk powders contain all twenty-one standard amino acids, the building blocks of proteins, and are high in soluble vitamins and minerals. According to USAID, the typical average amounts of major nutrients in the unreconstituted nonfat dry milk are (by weight) 36% protein, 52% carbohydrates (predominantly lactose), calcium 1.3%, potassium 1.8%. Whole milk powder, on the other hand, contains on average 25-27% protein, 36-38% carbohydrates, 26-40% fat, and 5-7% ash (minerals). In Canada, powdered milk must contain added vitamin D in an amount such that a reasonable daily intake of the milk will provide between 300 to 400 International Units (IU) of vitamin D. However, inappropriate storage conditions such as high relative humidity and high ambient temperature can significantly degrade the nutritive value of milk powder.

Commercial milk powders are reported to contain oxysterols (oxidized cholesterol) in higher amounts than in fresh milk (up to 30 µg/g, versus trace amounts in fresh milk). Oxysterols are derivatives of cholesterol that are produced either by free radicals or by enzymes. Some free radicals-derived oxysterols have been suspected of being initiators of atherosclerotic plaques. For comparison, powdered eggs contain even more oxysterols, up to 200 µg/g.

Export Market

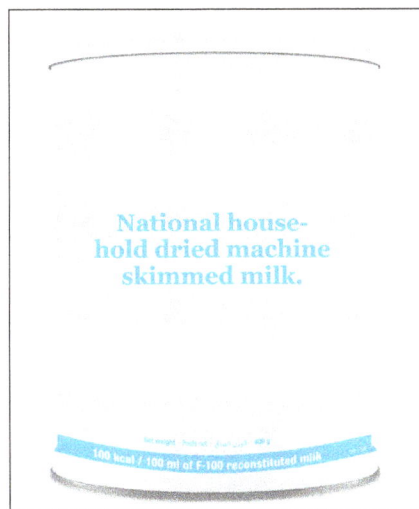

National household dried machine skimmed milk.

100 kcal / 100 ml of F-100 reconstituted milk

European production of milk powder is estimated around 800,000 tons of which the main volume is exported in bulk packing or consumer packs.

Brands on the market include "Nido", from the company Nestlé, "Incolac" from the company Milcobel, and "Dutch Lady" from FrieslandCampina.

Adulteration

In the Chinese milk scandal, melamine adulterant was found in Sanlu infant formula, added to fool tests into reporting higher protein content. Thousands became ill, and some children died, after consuming the product.

Contamination Scare

In August 2013, China temporarily suspended all milk powder imports from New Zealand, after a scare where botulism-causing bacteria was falsely detected in several batches of New Zealand-produced whey protein concentrate. As a result of the product recall, the New Zealand dollar slipped significantly based on expected losses in sales from this single commodity.

Use in Biotechnology

Fat-free powdered milk is used as a saturating agent to block nonspecific binding sites on supports like blotting membranes (nitrocellulose, polyvinylidene fluoride (PVDF) or nylon), preventing binding of further detection reagents and subsequent background. It may be referred as Blotto. The major protein of milk, casein, is responsible for most of the binding site saturation effect.

CREAM

A bottle of unhomogenised milk, with the cream
clearly visible, resting on top of the milk.

Cream is a dairy product composed of the higher-fat layer skimmed from the top of milk before homogenization. In un-homogenized milk, the fat, which is less dense, eventually rises to the top. In the industrial production of cream, this process is accelerated by using centrifuges called "separators". In many countries, it is sold in several grades depending on the total butterfat content. It can be dried to a powder for shipment to distant markets, and contains high levels of saturated fat.

Cream skimmed from milk may be called "sweet cream" to distinguish it from cream skimmed from whey, a by-product of cheese-making. Whey cream has a lower fat content and tastes more salty, tangy and "cheesy". In many countries, cream is usually sold partially fermented: sour cream, crème fraîche, and so on. Both forms have many culinary uses in sweet, bitter, salty and tangy dishes.

Cream produced by cattle (particularly Jersey cattle) grazing on natural pasture often contains some natural carotenoid pigments derived from the plants they eat; this gives it a slightly yellow tone, hence the name of the yellowish-white color: cream. This is also the origin of butter's yellow color. Cream from goat's milk, or from cows fed indoors on grain or grain-based pellets, is white.

Cream is used as an ingredient in many foods, including ice cream, many sauces, soups, stews, puddings, and some custard bases, and is also used for cakes. Whipped cream is served as a topping on ice cream sundaes, milkshakes, lassi, eggnog, sweet pies, strawberries, blueberries or peaches. Irish cream is an alcoholic liqueur which blends cream with whiskey, and often honey, wine, or coffee. Cream is also used in Indian curries such as masala dishes.

A slice of pumpkin pie topped with a whipped cream rose.

Cream (usually light/single cream or half and half) is often added to coffee in the US and Canada.

Both single and double cream can be used in cooking. Double cream or full-fat crème fraîche are often used when cream is added to a hot sauce, to prevent any problem with it separating or "splitting". Double cream can be thinned with milk to make an approximation of single cream.

The French word *crème* denotes not only dairy cream, but also other thick liquids such as sweet and savory custards, which are normally made with milk, not cream.

Processing and Additives

Cream may have thickening agents and stabilizers added. Thickeners include sodium alginate, carrageenan, gelatine, sodium bicarbonate, tetrasodium pyrophosphate, and alginic acid.

Other processing may be carried out. For example, cream has a tendency to produce oily globules (called "feathering") when added to coffee. The stability of the cream may be increased by increasing the non-fat solids content, which can be done by partial demineralisation and addition of sodium caseinate, although this is expensive.

Other Cream Products

Butter is made by churning cream to separate the butterfat and buttermilk. This can be done by hand or by machine.

Whipped cream is made by whisking or mixing air into cream with more than 30% fat, to turn the liquid cream into a soft solid. Nitrous oxide, from whipped-cream chargers may also be used to make whipped cream.

Sour cream, common in many countries including the U.S., Canada and Australia, is cream (12 to 16% or more milk fat) that has been subjected to a bacterial culture that produces lactic acid (0.5%+), which sours and thickens it.

Crème fraîche (28% milk fat) is slightly soured with bacterial culture, but not as sour or as thick as sour cream. Mexican crema (or cream espesa) is similar to crème fraîche.

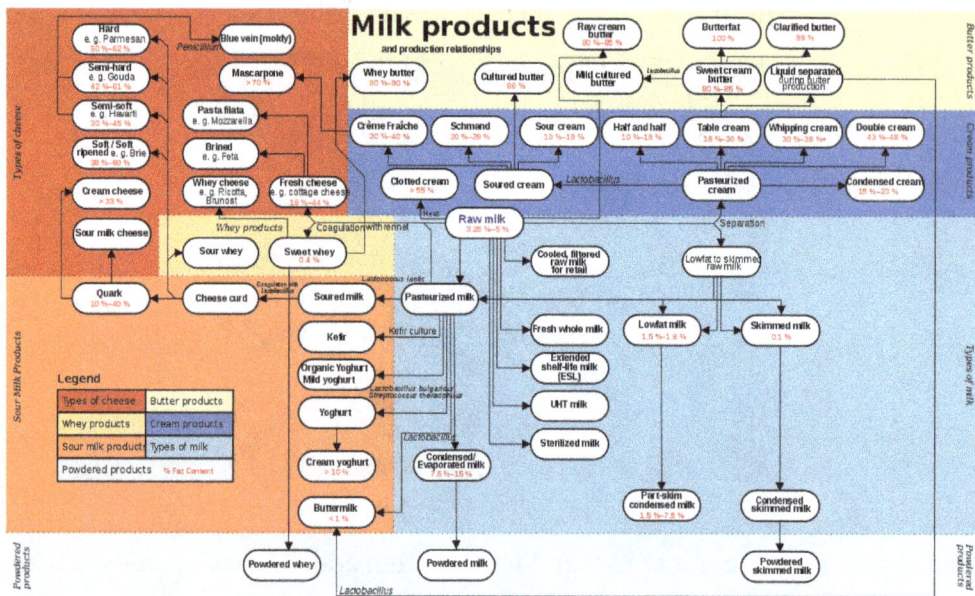

Chart of 50 types of milk products and relationships,
including cream (click on image to enlarge).

- Smetana is a heavy cream derived (15–40% milk fat) Central and Eastern European sweet or sour cream.

- Rjome or rømme is Norwegian sour cream containing 35% milk fat, similar to Icelandic sýrður rjómi.

- Clotted cream, common in the United Kingdom, is made through a process that starts by slowly heating whole milk to produce a very high-fat (55%) product. This is similar to Indian malai.

- Reduced cream is a cream product used in New Zealand to make Kiwi dip.

Other Items Called "Cream"

Some non-edible substances are called creams due to their consistency: shoe cream is runny, unlike regular waxy shoe polish; hand/body 'creme' or "skin cream" is meant for moisturizing the skin.

Regulations in many jurisdictions restrict the use of the word cream for foods. Words such as creme, kreme, creame, or whipped topping (e.g., Cool Whip) are often used for products which cannot legally be called cream, though in some jurisdictions even these spellings may be disallowed, for example under the doctrine of idem sonans. Oreo and Hydrox cookies are a type of sandwich cookie in which two biscuits have a soft, sweet filling between them that is called "crème filling." In some cases, foods can be described as cream although they do not contain predominantly milk fats; for example in Britain "ice cream" does not have to be a dairy product (although it must be labelled "contains non-milk fat"), and salad cream is the customary name for a condiment that has been produced since the 1920s.

In other languages, cognates of "cream" are also sometimes used for non-food products, such as fogkrém (Hungarian for toothpaste), or Sonnencreme (German for suntan lotion).

Types and Processes

Heat treatment (sterilisation, UHT, pasteurisation or thermisation), fat content, viscosity (liquid, semi-thick or thick), structure (whipped or whipping) and packing methods (aseptic or not, jars, pouches, bottles, bricks) are as many elements that help distinguish the numerous existing creams. These criteria may be combined, which gives, as a result, a broad range of creams:

* Raw cream: undergoes no heat treatment. This cream is neither pasteurised, nor sterilised. It comes directly from skimming. The reference "raw" has to be written on the label.

* Pasteurised cream: undergoes a heat treatment up to 72 °C during 20 seconds. This kind of product is nevertheless more fragile than sterilised cream.

* UHT cream: such cream is obtained from raw cream with a specific treatment. The Ultra High Temperature process means heating the cream at 145-150 °C during 2 seconds before quickly cooling it. This heat treatment does not alter any of the nutritional, tasting and functional qualities of the cream. It is always better to refrigerate UHT creams before distributing them in order to improve their stability during storage. Stabilising agents (carrageenan) may be added in order to stabilise the product even more. Such agents improve emulsion stability during sterilisation and prevent creaming during storage. Finally it helps reaching the expected texture. Carrageenan is a seaweed that grows mainly on the continental slope off Brittany. It is dried and ground into a powder to be added to food in small quantity. It is a polysaccharide from red seaweed that may be used as thickening agent and stabilising agent in food industry.

* Liquid cream: liquid and sweet, this cream has not been sown, which means there has been no addition of lactic ferments.

* Thick cream: has been matured, meaning that the cream has been sown with lactic ferments. Cream matures in tanks during several hours in order to acidify, thicken and intensify its taste.

* Crème fraîche: the designation "crème fraîche" is regulated by the decree from April 23rd 1980. In order to be given the designation "crème fraîche", the cream must comply with the following conditions: "undergo no sanitation heat treatment other than pasteurisation, be packed on the production location in the 24 hours following production". As a result sterilised cream can't obtain the designation "crème fraîche".

* Whipped cream: overrun emulsion obtained by introduction of air in the cream with an appropriate overrun at low-temperature (between 5 and 10 °C for a 30% fat cream). To get a stable foam, the cream must be cold-stored for several hours (usually 24 hours) before overrun. This low-temperature maturation creates a partial fat-crystallisation: crucial step to obtain a foam with the right texture and stability. A compromise is indeed looked for between liquid fat release, gathering and partial coalescing of fat globules around air bubbles.

- Crème Chantilly: The designation "Crème Chantilly" may only be used for whipped creams that have at least 30 grams of fat for 100 grams and in which no substance other than saccharose (semi-white, white or refined white sugar) or natural flavours have been added.

- AOP creams: two types of cream with a protected designation of origin exist:

  ○ Isigny cream, guaranteed without additives or stabilisers, it is a thick (sown with lactic ferments) and pasteurised cream. This cream undergoes a traditional maturation, it is left at room temperature for 16 to 18 hours. This maturation develops in the cream a sweet and slightly acid taste. This excellent product has been rewarded as early as 1986 with a Designation Protection of Origin (AOP).

  ○ Bresse cream, exists in semi-thick or thick. Both creams have been rewarded with an AOP.

- Acid cream: obtained from bacterial fermentation that produces lactic acid. This cream is broadly used in Anglo-Saxon countries, where it is called "sour cream".

Cream Manufacturing Processes

5 main steps:

- Skimming, centrifugation: separation of fat globules in milk. Milk skimming is done in a centrifugal cream separator.

- Fat standardisation: in order to obtain the expected fat content.

- Homogenisation: to prevent the creaming phenomenon during storage and to allow an increase in cream viscosity (for low-fat fluid creams).

- Heat treatment: the objective is to inactivate microbial lipases and as a result destroy pathogenic germs without damaging the cream organoleptic qualities. Most creams are pasteurised.

- Seeding and maturation: pasteurised creams may be matured with acidifying, aromatic or even thickening mesophilic lactic bacteria. Maturation gives more taste to the creams and protects it against lactic acid and bacteriocin production.

Main Industrial uses of Cream

Cream is used in many areas of the food industry, such as:

- Chocolate manufacturing: cream is used in some formulations, such as chocolate stuffing.

- Bakery, Pastry and Viennese Pastry: cream is used for stuffing in pastries, mostly for its taste and emulsifying capacities (in particular in whipped cream and mousse).

- Fresh dairy products: cream is used in most products: cream dessert, dairy dessert, fresh cheese spread, ice cream. Cream fat gives a soft texture and more flavour.

- Biscuit manufacturing: cream may be used for stuffing or for the biscuit itself.

- Ready meals, soups and broths: cream is used for its taste and binding property.

Functional Properties of Cream and its Advantages According to Industrial Applications

- Taste sensations: taste and flavour enhancer:
    - Maturation gives more flavour to the cream (transforming citrate into diacetyl thanks to Leuconostoc),
    - Size of suspended fat globules allows a fast fat melt.
- Texture builder: rich and velvety viscosity due to cream homogenisation, that is perfect for soups and sauces.
- Emulsifying property: cream proteins ease emulsification, ventilation, foaming and overrun.
- Whitening property:
    - Whitening effect,
    - Coloring effect due to fat globules and caseins that diffuse light.
- Browning of cooked aliments: Maillard reaction between proteins and lactose in the cream.

Sour Cream

Bowl of chili with sour cream and cheese.

Crisp potato skins with sour cream and chili sauce.

Mixed berries with sour cream and brown sugar.

Sour cream or soured cream is a dairy product obtained by fermenting regular cream with certain kinds of lactic acid bacteria. The bacterial culture, which is introduced either deliberately or naturally, sours and thickens the cream. Its name comes from the production of lactic acid by bacterial fermentation, which is called souring.

Traditional

Traditionally, sour cream was made by letting cream that was skimmed off the top of milk ferment at a moderate temperature. It can also be prepared by the souring of pasteurized cream with acid-producing bacterial culture. The bacteria that developed during fermentation thickened the cream and made it more acidic, a natural way of preserving it.

Commercial Varieties

Commercially produced sour cream usually contains not less than 14 percent milk fat. It may also contain milk and whey solids, buttermilk, starch in an amount not exceeding one percent, salt, and rennet derived from aqueous extracts from the fourth stomach of calves, kids or lambs, in an amount consistent with good manufacturing practice. In addition, according to the Canadian food regulations, the emulsifying, gelling, stabilizing and thickening agents in sour cream are algin, carob bean gum (locust bean gum), carrageenan, gelatin, guar gum, pectin, or propylene glycol alginate or any combination thereof in an amount not exceeding 0.5 per cent, monoglycerides, mono- and diglycerides, or any combination thereof, in an amount not exceeding 0.3 per cent, and sodium phosphate dibasic in an amount not exceeding 0.05 per cent.

Sour cream is not fully fermented, and like many dairy products, must be refrigerated unopened and after use. Additionally, in Canadian regulations, a milk coagulating enzyme derived from Rhizomucor miehei (Cooney and Emerson) from Mucor pusillus Lindt by pure culture fermentation process or from Aspergillus oryzae RET-1 (pBoel777) can also be added into sour cream production process, in an amount consistent with good manufacturing practice. Sour cream is sold with an expiration date stamped on the container, though whether this is a "sell by", a "best by" or a "use by" date varies with local regulation. Refrigerated unopened sour cream can last for 1–2 weeks beyond its sell by date while refrigerated opened sour cream generally lasts for 7–10 days.

Physical-Chemical Properties

Ingredients

Cultured cream: Processed sour cream can include any of the following additives and preservatives: grade A whey, modified food starch, sodium phosphate, sodium citrate, guar gum, carrageenan, calcium sulfate, potassium sorbate, and locust bean gum.

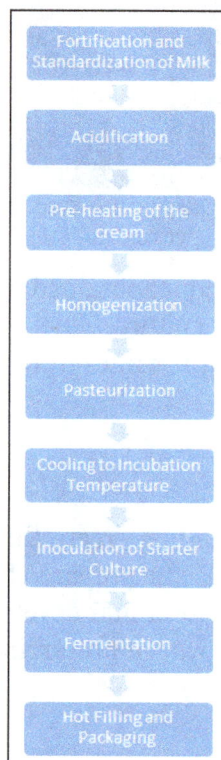

Fortification and Standardization of Milk

↓

Acidification

↓

Pre-heating of the cream

↓

Homogenization

↓

Pasteurization

↓

Cooling to Incubation Temperature

↓

Inoculation of Starter Culture

↓

Fermentation

↓

Hot Filling and Packaging

Simple illustration of the processing
order of sour cream manufacturing.

Protein Composition

Milk is made up of approximately 3.0-3.5% protein. The main proteins in cream are caseins and whey proteins. Of the total fraction of milk proteins, caseins make up 80% while the whey proteins make up 20%. There are four main classes of caseins; β-caseins, α(s1)-caseins, α(s2)-casein and κ-caseins. These casein proteins form a multi molecular colloidal particle known as a casein micelle. The proteins mentioned have an affinity to bind with other casein proteins, or to bind with calcium phosphate, and this binding is what forms the aggregates. The casein micelles are aggregates of β-caseins, α(s1)-caseins, α(s2)-caseins, that are coated with κ-caseins. The proteins are held together by small clusters of calcium phosphate, the micelle also contains lipase, citrate, minor ions, and plasmin enzymes, along with entrapped milk serum. Casein micelles are rather porous structures, ranging in the size of 50-250 nm in diameter and the structures on average are 6-12% of the total volume fraction of milk. The structure is porous in order to be able to hold a sufficient amount of water, its structure also assists in the reactivity of the micelle. The formation of casein molecules into the micelle is very unusual due to β-casein's large amount of propyl residues (the proline residues disrupt the formation of α-helixes and β-sheets) and because κ-caseins only contain one phosphorylation residue (they are glycoproteins). Due to κ-caseins being glycoproteins, they are stable in the presence of calcium ions so the κ-caseins are on the outer layer of the micelle to partially protect the non glycoproteins β-caseins, α(s1)-caseins, α(s2)-caseins from precipitating out in the presence of excess calcium ions. Casein micelles are not heat sensitive particles, they are pH sensitive. The colloidal particles are stable at the normal pH of milk which is 6.5-6.7, the micelles will precipitate at the isoelectric point of milk which is a pH of 4.6.

The proteins that make up the remaining 20% of the fraction of proteins in cream are known as whey proteins. Whey proteins are also widely referred to as serum proteins, which is used when the casein proteins have been precipitated out of solution. The two main components of whey proteins in milk are β-lactoglobulin and α-lactalbumin. The remaining whey proteins in milk are; immunoglobulins, bovine serum albumin, and enzymes such as lysozyme. Whey proteins are much more water-soluble than casein proteins. The main biological function of β-lactoglobulin in milk is to serve as a way to transfer vitamin A, and the main biological function of α-lactalbumin in lactose synthesis. The whey proteins are very resistant to acids and proteolytic enzymes. However whey proteins are heat sensitive proteins, the heating of milk will cause the denaturation of the whey proteins. The denaturation of these proteins happens in two steps. The structures of β-lactoglobulin and α-lactalbumin unfold, and then the second step is the aggregation of the proteins within milk. This is one of the main factors that allows whey proteins to have such good emulsifying properties. Native whey proteins are also known for their good whipping properties, and in milk products as described above their gelling properties. Upon denaturation of whey proteins, there is an increase in the water holding capacity of the product.

Processing

The manufacturing of sour cream begins with the standardization of fat content; this step is to ensure that the desired or legal amount of milk fat is present. As previously mentioned the minimum amount of milk fat that must be present in sour cream is 18%. During this step in the manufacturing process other dry ingredients are added to the cream; additional grade A whey for example would be added at this time. Another additive used during this processing step are a series of ingredients known as stabilizers. The common stabilizers that are added to sour cream are polysaccharides and gelatin, including modified food starch, guar gum, and carrageenans. The reasoning behind the addition of stabilizers to fermented dairy products is to provide smoothness in the body and texture of the product. The stabilizers also assist in the gel structure of the product and reduce whey syneresis. The formation of these gel structures, leaves less free water for whey syneresis, thereby extending the shelf life. Whey syneresis is the loss of moisture by the expulsion of whey. This expulsion of whey can occur during the transportation of containers holding the sour cream, due to the susceptibility to motion and agitation. The next step in the manufacturing process is the acidification of the cream. Organic acids such as citric acid or sodium citrate are added to the cream prior to homogenization in order to increase the metabolic activity of the starter culture. To prepare the mixture for homogenization, it is heated for a short period of time.

Homogenization is a processing method that is utilized to improve the quality of the sour cream in regards to the color, consistency, creaming stability, and creaminess of the cultured cream. During homogenization larger fat globules within the cream are broken down into smaller sized globules to allow an even suspension within the system. At this point in the processing the milk fat globules and the casein proteins are not interacting with each other, there is repulsion occurring. The mixture is homogenized, under high pressure homogenization above 130 bar (unit) and at a high temperature of 60 °C. The formation of the small globules (below 2 microns in size) previously mentioned allows for reducing a cream layer formation and increases the viscosity of the product. There is also a reduction in the separation of whey, enhancing the white color of the sour cream.

After homogenization of the cream, the mixture must undergo pasteurization. Pasteurization is a mild heat treatment of the cream, with the purpose of killing any harmful bacteria in the cream. The homogenized cream undergoes high temperature short time (HTST) pasteurization method. In this type of pasteurization the cream is heated to the high temperature of 85 °C for thirty minutes. This processing step allows for a sterile medium for when it is time to introduce the starter bacteria.

After the process of pasteurization, there is a cooling process where the mixture is cooled down to a temperature of 20°C. The reason that the mixture was cooled down to the temperature of 20 °C is due to the fact that this is an ideal temperature for mesophilic inoculation. After the homogenized cream has been cooled to 20°C, it is inoculated with 1-2% active starter culture. The type of starter culture utilized is essential for the production of sour cream. The starter culture is responsible for initiating the fermentation process by enabling the homogenized cream to reach the pH of 4.5 to 4.8. Lactic acid bacteria (hereto known as LAB) ferment lactose to lactic acid, they are Gram-positive facultative anaerobes. The strains of LAB that are utilized to allow the fermentation of sour cream production are Lactococcus lactis subsp latic or Lactococcus lactis subsp cremoris they are lactic acid bacteria associated with producing the acid. The LAB that are known for producing the aromas in sour cream are Lactococcus lactis ssp. lactis biovar diacetyllactis. Together these bacteria produce compounds that will lower the pH of the mixture, and produce flavor compounds such as diacetyl.

After the inoculation of starter culture, the cream is portioned in packages. For 18 hours a fermentation process takes place in which the pH is lowered from 6.5 to 4.6. After fermentation, one more cooling process takes place. After this cooling process the sour cream is packaged into their final containers and sent to the market.

Physical-chemical Changes

Sour cream can also be fried in oil or fat, and used
on top of noodle dishes, as in Hungarian cuisine.

During the pasteurization process, temperatures are raised past the point where all the particles in the system are stable. When cream is heated to temperatures above 70 °C, there is denaturation of whey proteins. To avoid phase separation brought on by the increased surface area, the fat globules readily bind with the denatured β-lactoglobulin. The absorption of the denatured whey proteins (and whey proteins that bound with casein micelles) increases the number of structural components in the product; the texture of sour cream can be partly attributed to this. The denaturation of whey proteins is also known for increasing the strength of the cross-linking within the cream system, due to the formation of whey protein polymers.

When the cream is inoculated with starter bacteria and the bacteria begins converting lactose to lactic acid, the pH begins a slow decrease. When this decrease begins, dissolution of calcium phosphate occurs, and causes a rapid drop in the pH. During the processing step fermentation the pH was dropped from 6.5 to 4.6, this drop in pH brings on a physicochemical change to the casein micelles. Recall the casein proteins are heat stable, but they are not stable in certain acidic conditions. The colloidal particles are stable at the normal pH of milk which is 6.5-6.7, the micelles will precipitate at the isoelectric point of milk which is a pH of 4.6. At a pH of 6.5 the casein micelles repulse each other due to the electronegativity of the outer layer of the micelle. During this drop in pH there is a reduction in zeta potential, from the highly net negative charges in cream to no net charge when approaching the PI.

$$U_E = \left| \frac{2\varepsilon z f(ka)}{3\eta} \right|$$

The formula shown is the Henry's equation, where z: zeta potential, U_E: electrophoretic mobility, ε: dielectric constant, η: viscosity, and f(Ka): Henry's function. This equation is used to find the zeta potential, which is calculated to find the electrokinetic potential in colloidal dispersions. Through electrostatic interactions the casein molecules begin approaching and aggregating together. The casein proteins enter a more ordered system, attributing to a strong gel structure formation. The whey proteins that were denatured in the heating steps of processing, are insoluble at this acidic pH and are precipitated with casein.

The interactions involved in gelation and aggregation of casein micelles are hydrogen bonds, hydrophobic interactions, electrostatic attractions and vander waals attractions. These interactions are highly dependent on pH, temperature and time. At the isoelectric point, the net surface charge of casein micelle is zero and a minimum of electrostatic repulsion can be expected. Furthermore, aggregation is taking place due to dominating hydrophobic interactions. Differences in the zeta potential of milk can be caused by differences in ionic strength differences, which in turn depend on the amount of calcium present in the milk. The stability of milk is largely due to the electrostatic repulsion of casein micelles. These casein micelles aggregated and precipitated when they approach the absolute zeta potential values at pH 4.0 – 4.5. When the heat treated and denatured, whey protein is covering the casein micelle, isoelectric point of the micelle elevated to the isoelectric point of β lactoglobulin (approximately pH 5.3).

Rheological Properties

Sour cream exhibits time-dependent thixotropic behaviors. Thixotropic fluids reduce in viscosity as work is applied, and when the product is no longer under stress, the fluid returns to its previous viscosity. The viscosity of sour cream at room temperature is 100,000 cP, (for comparison: water has a viscosity of 1 cP at 20 °C). The thixotropic properties exhibited by sour cream are what make it such a versatile product in the food industry.

Whipped Cream

Whipped cream is cream that is whipped by a whisk or mixer until it is light and fluffy. Whipped cream is often sweetened and sometimes flavored with vanilla. Whipped cream is also called Chantilly cream or Crème chantilly.

Food Chemistry

Whipped cream is a culinary colloid produced when heavy cream is subjected to mechanical aeration. Air is incorporated into cream containing at least 35% fat by one of two processes: mechanical agitation with a high-speed blade or whip, or injecting a gas under high pressure, which expands rapidly when released from pressurized containment.

During whipping, partially coalesced fat molecules create a stabilized network which traps air bubbles. The resulting colloid is roughly double the volume of the original cream. If, however, the whipping is continued, the fat droplets will stick together destroying the colloid and forming butter. Lower-fat cream (or milk) does not whip well, while higher-fat cream produces a more stable foam.

Methods of Whipping

Cream is usually whipped with a whisk, an electric or hand mixer, or a food processor. Results are best when the equipment and ingredients are cold.

Whipped cream is often flavored with sugar, vanilla, coffee, chocolate, orange, and so on. Many 19th-century recipes recommend adding gum tragacanth to stabilize whipped cream; a few include whipped egg whites. Various other substances, including gelatin and diphosphate (E450), are used in commercial stabilizers.

Whipped cream may also be made in a whipping siphon, typically using nitrous oxide as the gas, as carbon dioxide tends to give a sour taste (*cf.* soda syphon). The siphon may have replaceable cartridges or be sold as a pre-pressurized retail package. The gas dissolves in the butterfat under pressure, and when the pressure is released, produces bubbles and thus whipped cream.

Crème Chantilly

Crème Chantilly is another name for whipped cream. The difference between "whipped cream" and "crème Chantilly" is not systematic. Some authors distinguish between the two, with crème Chantilly being sweetened, and whipped cream not. However, most authors treat the two as synonyms, with both being sweetened, neither being sweetened, or treating sweetening as optional. Many authors use only one of the two names (for the sweetened or unsweetened version), so it is not clear if they distinguish the two.

The invention of crème Chantilly is often credited incorrectly, and without evidence, to François Vatel, maître d'hôtel at the Château de Chantilly in the mid-17th century. But the name Chantilly is first connected with whipped cream in the mid-18th century, around the time that the Baronne d'Oberkirch praised the "cream" served at a lunch at the Hameau de Chantilly—but did not say what exactly it was, or call it Chantilly cream.

The names "crème Chantilly", "crème de Chantilly", "crème à la Chantilly", or "crème fouettée à la Chantilly" only become common in the 19th century. In 1806, the first edition of Viard's Cuisinier Impérial mentions neither "whipped" nor "Chantilly" cream, but the 1820 edition mentions both.

The name Chantilly was probably used because the château had become a symbol of refined food.

Imitation Whipped Cream

Imitations of whipped cream, often sold under the name whipped topping or squirty cream, are commercially available. They may be used for various reasons:

- To exclude dairy ingredients.
  - For milk allergies.
  - For vegan diets.
  - For religious reasons, such as dietary laws forbidding mixing meat with dairy.
- To provide extended shelf life (often in the freezer).
- To reduce the price although some popular brands cost twice as much as whipped cream.
- For convenience.

Artificial whipped cream normally contains some mixture of partially hydrogenated oil, sweeteners, water, and stabilizers and emulsifiers added to prevent syneresis, similar to margarine instead of the butter fat in the cream used in whipped cream.

It may be sold frozen in plastic tubs (*e.g.*, Cool Whip) or in aerosol containers (*e.g.* Reddi-wip), or in liquid form in cartons, reminiscent of real whipping cream.

Uses

Whipped cream or crème Chantilly is a popular topping for fruit and desserts such as pie, ice cream (especially sundaes), cupcakes, cakes, milkshakes, waffles, hot chocolate, cheesecakes, Jello and puddings. It is also served on coffee, especially in the Viennese coffee house tradition, where coffee with whipped cream is known as Melange mit Schlagobers. Whipped cream is used as an ingredient in many desserts, for example as a filling for profiteroles and layer cakes.

It is often piped onto a dish using a pastry bag to create decorative shapes.

The structure of whipped cream is very similar to the fat and air structure that exists in ice cream. Cream is an emulsion with a fat content of 35-40%. When you whip a bowl of heavy cream, the agitation and the air bubbles that are added cause the fat globules to begin to partially coalesce in chains and clusters and adsorb to and spread around the air bubbles.

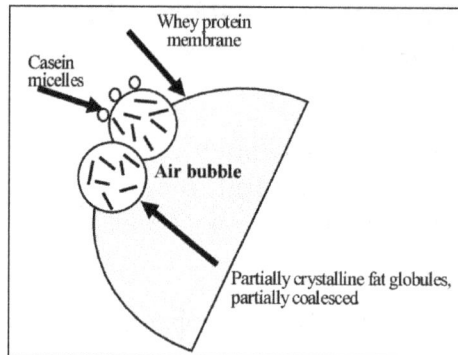

As the fat partially coalesces, it causes one fat-stabilized air bubble to be linked to the next, and so on. The whipped cream soon starts to become stiff and dry appearing and takes on a smooth texture. This results from the formation of this partially coalesced fat structure stabilizing the air bubbles. The water, lactose and proteins are trapped in the spaces around the fat-stabilized air bubbles. The crystalline fat content is essential (hence whipping of cream is very temperature dependent) so that the fat globules partially coalesce into a 3-dimensional structure rather than fully coalesce into larger and larger globules that are not capable of structure-building. This is caused by the crystals within the globules that cause them to stick together into chains and clusters, but still retain the individual identity of the globules. If whipped cream is whipped too far, the fat will begin to churn and butter particles will form.

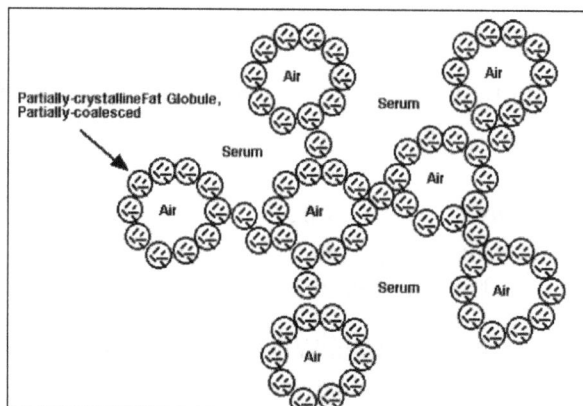

Below are scanning electron micrographs image of whipped cream. If you compare the schematics above with the "real thing" below, you should be able to fully understand whipped cream structure.

The structure of whipped cream as determined by scanning electron microscopy. A. the relative size and prevalence of air bubbles (a) and fat globules (f); bar = 30 um. B. Internal structure of the air bubble, showing the layer of partially coalesced fat which has stabilized the bubble; bar = 5 um. C. Details of the partially coalesced fat layer, showing the interaction of the individual fat globules. Bar = 3 um.

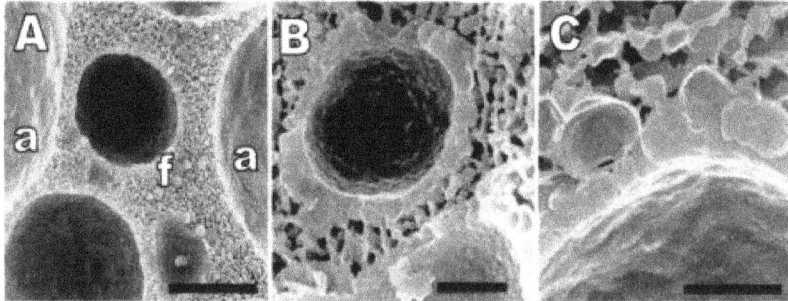

Fat partial coalescence as it affects things like whipped cream and ice cream structure is an active area of our research here at the University of Guelph.

Production of Whipped Cream

The production of whipping cream will be described first, followed by a short description of the production of whipped cream in aerosols.

Raw Materials

Whipping cream is made from whole milk. It is important that this milk contains little heat-resistant bacteria's, because the presence of Bacillus Cereus can lead to an instable fat emulsion. The formation of heat-resistant lipase due to psychrotroph bacteria's is also harmful and may not occur. To prevent auto-oxidation leading to spoiled taste, not even small amounts of cupper are allowed to be present in the milk.

Separating Cream

To obtain cream with the proper fat content, the cream is separated from the milk by means of a separator. This separating generally takes place at a temperature of approximately 50 °C, because a good separation of the fat takes place at this temperature. However, the cream can also be separated from cold milk. By separating at approximately 7 °C, a cream is obtained with good whipping characteristics, although the fat content of the milk obtained at this temperature is usually higher than normally.

Standardizing

The cream is standardized to a fat content of at least 35% by means of the previously obtained cream and milk. To lower the serum discharge of the whipped cream, skimmed milk powder or thickenings agents can be added to the standardized cream.

Pasteurizing

After the standardizing, the cream is pasteurized again. The time and temperature combination differs per concern, but a standpasteurization for 30 minutes at 80 °C or heating up too 100 °C in a heat exchanger is not uncommon. It is also possible to pasteurize the cream for 20 minutes at 103 °C while it is already wrapped.

When one wants to remove unwanted scents from the cream, one can choose to keep the cream moving at high temperature in an open kettle. The cream has to be kept moving very accurately, for clotting of the balls of fat needs to be prevented.

Cooling and Wrapping

The whole is cooled to 5 °C after pasteurization and then wrapped in glass bottles or in disposable wrappings.

Afterwards Pasteurizing/Sterilizing

To ensure the product keeps fresh for a longer period of time, the cream can be pasteurized afterwards or sterilized. When the cream is sterilized, it first needs to be homogenized lightly (at 1,5 bar), to prevent coalescence. However, due to the homogenizing, the whipping characteristics of the cream deteriorate. It is therefore sensible to add an emulsifier to the cream. After the final pasteurizing or sterilizing, the whole needs to be cooled quickly to at least 5 °C.

Rebodying

The whipping characteristics of the cream can be improved by keeping the cream, after cooling to 5 °C, at this temperature for several hours. The cream is then heated shortly to 20 to 30 °C and cooled again to 5 °C.

Crystallizing

Finally, the cream has to be kept at 4 °C for 24 hours, in order for the milk fat to crystallize, which improves the whipping characteristics.

Whipped Cream

The cream which is obtained by separation and standardisation is mixed in a mixing tank with emulsifiers and stabilizers. The whole is then sterilized to ensure a longer freshness. The sterilization process is the same for whipped cream as for whipping cream. After cooling the whole to 5 °C, the aerosols are aseptically filled with the cream. Before the aerosols are filled, they are brought to an overpressure cabin, where they are heated to 160 °C. When the aerosols are filled, a cap is placed on top of them. The propellant is then injected in the cream. The gas N2O is usually used, because it dissolves relatively good in fat. The propellant dissolves in the cream, but when the cream is pushed out of the aerosol under pressure, the gas leaves the solution again. The laughing gas therefore is used to push the cream out. However, the whipped cream will return to a more or less liquid state again after a short period of time, because the laughing gas diffuses out of the cream as it is not present in the atmosphere.

BUTTER

Butter is a dairy product with high butterfat content which is solid when chilled and at room temperature in some regions, and liquid when warmed. It is made by churning fresh or fermented cream or milk to separate the butterfat from the buttermilk. It is generally used as a spread on plain or toasted bread products and a condiment on cooked vegetables, as well as in cooking, such as baking, sauce making, and pan frying. Butter consists of butterfat, milk proteins and water, and often added salt.

Most frequently made from cow's milk, butter can also be manufactured from the milk of other mammals, including sheep, goats, buffalo, and yaks. Salt (such as dairy salt), flavorings (such as garlic) and preservatives are sometimes added to butter. Rendering butter, removing the water and milk solids, produces clarified butter or *ghee*, which is almost entirely butterfat.

Butter is a water-in-oil emulsion resulting from an inversion of the cream, where the milk proteins are the emulsifiers. Butter remains a firm solid when refrigerated, but softens to a spreadable consistency at room temperature, and melts to a thin liquid consistency at 32 to 35 °C (90 to 95 °F). The density of butter is 911 grams per litre (0.950 lb per US pint). It generally has a pale yellow color, but varies from deep yellow to nearly white. Its natural, unmodified color is dependent on the source animal's feed and genetics, but the commercial manufacturing process commonly manipulates the color with food colorings like annatto or carotene.

Production

Churning cream into butter using a hand-held mixer.

Unhomogenized milk and cream contain butterfat in microscopic globules. These globules are surrounded by membranes made of phospholipids (fatty acid emulsifiers) and proteins, which prevent the fat in milk from pooling together into a single mass. Butter is produced by agitating cream, which damages these membranes and allows the milk fats to conjoin, separating from the other parts of the cream. Variations in the production method will create butters with different consistencies, mostly due to the butterfat composition in the finished product. Butter contains fat in three separate forms: free butterfat, butterfat crystals, and undamaged fat globules. In the finished product, different proportions of these forms result in different consistencies within the butter; butters with many crystals are harder than butters dominated by free fats.

Churning produces small butter grains floating in the water-based portion of the cream. This watery liquid is called buttermilk—although the buttermilk most common today is instead a directly fermented skimmed milk. The buttermilk is drained off; sometimes more buttermilk is removed by rinsing the grains with water. Then the grains are "worked": pressed and kneaded together. When prepared manually, this is done using wooden boards called scotch hands. This consolidates the butter into a solid mass and breaks up embedded pockets of buttermilk or water into tiny droplets.

Commercial butter is about 80% butterfat and 15% water; traditionally made butter may have as little as 65% fat and 30% water. Butterfat is a mixture of triglyceride, a triester derived from glycerol and three of any of several fatty acid groups. Butter becomes rancid when these chains break down into smaller components, like butyric acid and diacetyl. The density of butter is 0.911 g/cm^3 (0.527 oz/in^3), about the same as ice.

In some countries, butter is given a grade before commercial distribution.

Types

Before modern factory butter making, cream was usually collected from several milkings and was therefore several days old and somewhat fermented by the time it was made into butter. Butter made from a fermented cream is known as cultured butter. During fermentation, the cream naturally sours as bacteria convert milk sugars into lactic acid. The fermentation process produces additional aroma compounds, including diacetyl, which makes for a fuller-flavored and more "buttery" tasting product. Today, cultured butter is usually made from pasteurized cream whose fermentation is produced by the introduction of Lactococcus and Leuconostoc bacteria.

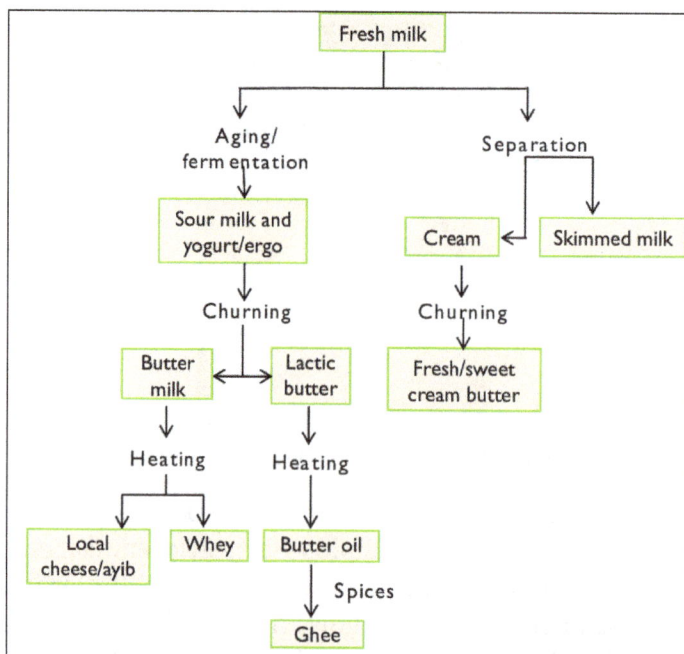

Chart of milk products and production relationships, including butter.

Another method for producing cultured butter, developed in the early 1970s, is to produce butter from fresh cream and then incorporate bacterial cultures and lactic acid. Using this method, the cultured butter flavor grows as the butter is aged in cold storage. For manufacturers, this method is more efficient, since aging the cream used to make butter takes significantly more space than simply storing the finished butter product. A method to make an artificial simulation of cultured butter is to add lactic acid and flavor compounds directly to the fresh-cream butter; while this more efficient process is claimed to simulate the taste of cultured butter, the product produced is not cultured but is instead flavored.

Dairy products are often pasteurized during production to kill pathogenic bacteria and other microbes. Butter made from pasteurized fresh cream is called sweet cream butter. Production of sweet cream butter first became common in the 19th century, with the development of refrigeration and the mechanical cream separator. Butter made from fresh or cultured unpasteurized cream is called raw cream butter. While butter made from pasteurized cream may keep for several months, raw cream butter has a shelf life of roughly ten days.

Throughout continental Europe, cultured butter is preferred, while sweet cream butter dominates in the United States and the United Kingdom. Cultured butter is sometimes labeled "European-style" butter in the United States, although cultured butter is made and sold by some, especially Amish, dairies. Commercial raw cream butter is virtually unheard-of in the United States. Raw cream butter is generally only found made at home by consumers who have purchased raw whole milk directly from dairy farmers, skimmed the cream themselves, and made butter with it. It is rare in Europe as well.

Several "spreadable" butters have been developed. These remain softer at colder temperatures and are therefore easier to use directly out of refrigeration. Some methods modify the makeup of the butter's fat through chemical manipulation of the finished product, some manipulate the cattle's feed, and some incorporate vegetable oil into the butter. "Whipped" butter, another product designed to be more spreadable, is aerated by incorporating nitrogen gas—normal air is not used to avoid oxidation and rancidity.

All categories of butter are sold in both salted and unsalted forms. Either granular salt or a strong brine are added to salted butter during processing. In addition to enhanced flavor, the addition of salt acts as a preservative. The amount of butterfat in the finished product is a vital aspect of production. In the United States, products sold as "butter" must contain at least 80% butterfat. In practice, most American butters contain slightly more than that, averaging around 81% butterfat. European butters generally have a higher ratio—up to 85%.

Liquid clarified butter.

Clarified butter is butter with almost all of its water and milk solids removed, leaving almost-pure butterfat. Clarified butter is made by heating butter to its melting point and then allowing it to cool; after settling, the remaining components separate by density. At the top, whey proteins form a skin, which is removed. The resulting butterfat is then poured off from the mixture of water and casein proteins that settle to the bottom.

Ghee is clarified butter that has been heated to around 120 °C (250 °F) after the water evaporated, turning the milk solids brown. This process flavors the ghee, and also produces antioxidants that help protect it from rancidity. Because of this, ghee can keep for six to eight months under normal conditions.

Whey Butter

Butter made in a barn.

Cream may be separated (usually by a centrifugal separator) from whey instead of milk, as a by-product of cheese making. Whey butter may be made from whey cream. Whey cream and butter have a lower fat content and taste more salty, tangy and "cheesy". They are also cheaper than "sweet" cream and butter. The fat content of whey is low, so 1000 pounds of whey will typically give 3 pounds of butter.

European Butters

There are several butters produced in Europe with protected geographical indications; these include:

- Beurre d'Ardenne, from Belgium.

- Beurre d'Isigny, from France.

- Beurre Charentes-Poitou (Which also includes: Beurre des Charentes and Beurre des Deux-Sèvres under the same classification), from France.

- Beurre Rose, from Luxembourg.

- Mantequilla de Soria, from Spain.

- Mantega de l'Alt Urgell i la Cerdanya, from Spain.

- Rucava white butter (*Rucavas baltais sviests*), from Latvia.

Packaging

Elsewhere

Outside of the United States, the shape of butter packages is approximately the same, but the butter is measured for sale and cooking by mass (rather than by volume or unit/stick), and is sold in 250 g (8.8 oz) and 500 g (18 oz) packages. The wrapper is usually a foil and waxed-paper laminate. The waxed paper is now a siliconised substitute, but is still referred to in some places as parchment, from the wrapping used in past centuries; and the term 'parchment-wrapped' is still employed where the paper alone is used, without the foil laminate.

Butter for commercial and industrial use is packaged in plastic buckets, tubs, or drums, in quantities and units suited to the local market.

Worldwide

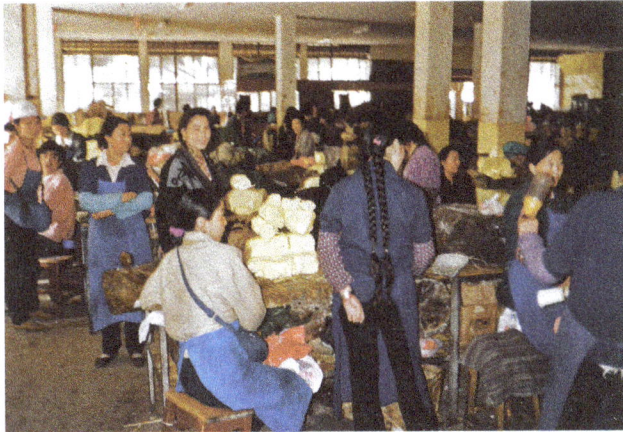
Butter market, Lhasa, Tibet.

In 1997, India produced 1,470,000 metric tons (1,620,000 short tons) of butter, most of which was consumed domestically. Second in production was the United States (522,000 t or 575,000 short tons), followed by France (466,000 t or 514,000 short tons), Germany (442,000 t or 487,000 short tons), and New Zealand (307,000 t or 338,000 short tons). France ranks first in per capita butter consumption with 8 kg per capita per year. In terms of absolute consumption, Germany was second after India, using 578,000 metric tons (637,000 short tons) of butter in 1997, followed by France (528,000 t or 582,000 short tons), Russia (514,000 t or 567,000 short tons), and the United States (505,000 t or 557,000 short tons). New Zealand, Australia, and the Ukraine are among the few nations that export a significant percentage of the butter they produce.

Different varieties are found around the world. Smen is a spiced Moroccan clarified butter, buried in the ground and aged for months or years. A similar product is maltash of the Hunza Valley, where cow and yak butter can be buried for decades, and is used at events such as weddings. Yak butter is a specialty in Tibet; tsampa, barley flour mixed with yak butter, is a staple food. Butter tea is consumed in the Himalayan regions of Tibet, Bhutan, Nepal and India. It consists of tea served with intensely flavored or "rancid" yak butter and salt. In African and Asian developing nations, butter is traditionally made from sour milk rather than cream. It can take several hours of churning to produce workable butter grains from fermented milk.

Storage and Cooking

Hollandaise sauce served over white asparagus and potatoes.

Normal butter softens to a spreadable consistency around 15 °C (60 °F), well above refrigerator temperatures. The butter compartment found in many refrigerators may be one of the warmer sections inside, but it still leaves butter quite hard. Until recently, many refrigerators sold in New Zealand featured a butter conditioner, a compartment kept warmer than the rest of the refrigerator but still cooler than room temperature with a small heater. Keeping butter tightly wrapped delays rancidity, which is hastened by exposure to light or air, and also helps prevent it from picking up other odors. Wrapped butter has a shelf life of several months at refrigerator temperatures. Butter can also be frozen to further extend its storage life.

"French butter dishes" or "Acadian butter dishes" have a lid with a long interior lip, which sits in a container holding a small amount of water. Usually the dish holds just enough water to submerge the interior lip when the dish is closed. Butter is packed into the lid. The water acts as a seal to keep the butter fresh, and also keeps the butter from overheating in hot temperatures. This method lets butter sit on a countertop for several days without spoiling.

Once butter is softened, spices, herbs, or other flavoring agents can be mixed into it, producing what is called a compound butter or composite butter (sometimes also called composed butter). Compound butters can be used as spreads, or cooled, sliced, and placed onto hot food to melt into a sauce. Sweetened compound butters can be served with desserts; such hard sauces are often flavored with spirits.

When heated, butter quickly melts into a thin liquid.

Melted butter plays an important role in the preparation of sauces, most obviously in French cuisine. Beurre noisette (hazelnut butter) and Beurre noir (black butter) are sauces of melted butter cooked until the milk solids and sugars have turned golden or dark brown; they are often finished with an addition of vinegar or lemon juice. Hollandaise and béarnaise sauces are emulsions of egg yolk and melted butter; they are in essence mayonnaises made with butter instead of oil. Hollandaise and béarnaise sauces are stabilized with the powerful emulsifiers in the egg yolks, but butter itself contains enough emulsifiers mostly remnants of the fat globule membranes to form a stable emulsion on its own. Beurre blanc (white butter) is made by whisking butter into reduced vinegar or wine, forming an emulsion with the texture of thick cream. Beurre monté (prepared butter) is melted but still emulsified butter; it lends its name to the practice of "mounting" a sauce with butter: whisking cold butter into any water-based sauce at the end of cooking, giving the sauce a thicker body and a glossy shine as well as a buttery taste.

In Poland, the butter lamb (Baranek wielkanocny) is a traditional addition to the Easter Meal for many Polish Catholics. Butter is shaped into a lamb either by hand or in a lamb-shaped mould. Butter is also used to make edible decorations to garnish other dishes.

Mixing melted butter with chocolate to make a brownie.

Butter is used for sautéing and frying, although its milk solids brown and burn above 150 °C (250 °F) a rather low temperature for most applications. The smoke point of butterfat is around 200 °C (400 °F), so clarified butter or ghee is better suited to frying. Ghee has always been a common frying medium in India, where many avoid other animal fats for cultural or religious reasons.

Butter fills several roles in baking, where it is used in a similar manner as other solid fats like lard, suet, or shortening, but has a flavor that may better complement sweet baked goods. Many cookie doughs and some cake batters are leavened, at least in part, by creaming butter and sugar together, which introduces air bubbles into the butter. The tiny bubbles locked within the butter expand in the heat of baking and aerate the cookie or cake. Some cookies like shortbread may have no other source of moisture but the water in the butter. Pastries like pie dough incorporate pieces of solid fat into the dough, which become flat layers of fat when the dough is rolled out. During baking, the fat melts away, leaving a flaky texture. Butter, because of its flavor, is a common choice for the fat in such a dough, but it can be more difficult to work with than shortening because of its low melting point. Pastry makers often chill all their ingredients and utensils while working with a butter dough.

Butter is essentially the fat of the milk. It is usually made from sweet cream and is salted. However, it can also be made from acidulated or bacteriologically soured cream and saltless (sweet) butters are also available. Well into the 19th century butter was still made from cream that had been allowed to stand and sour naturally. The cream was then skimmed from the top of the milk and poured into a wooden tub. Buttermaking was done by hand in butter churns. The natural souring process is, however, a very sensitive one and infection by foreign micro-organisms often spoiled the result. Today's commercial buttermaking is a product of the knowledge and experience gained over the years in such matters as hygiene, bacterial acidifying and heat treatment, as well as the rapid technical development that has led to the advanced machinery now used. The commercial cream separator was introduced at the end of the 19th century, the continuous churn had been commercialized by the middle of the 20th century.

Buttermaking Process

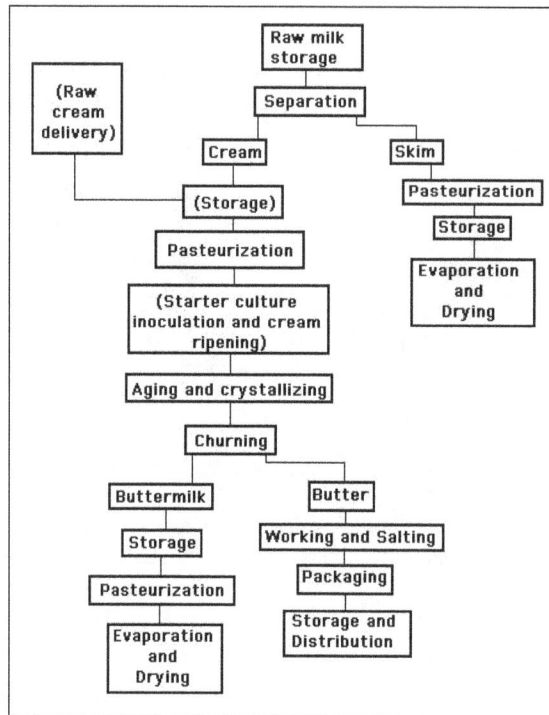

The buttermaking process involves quite a number of stages. The continuous buttermaker has become the most common type of equipment used.

The cream can be either supplied by a fluid milk dairy or separated from whole milk by the butter manufacturer. The cream should be sweet (pH >6.6, TA = 0.10 - 0.12%), not rancid and not oxidized.

If the cream is separated by the butter manufacturer, the whole milk is preheated to the required temperature in a milk pasteurizer before being passed through a separator. The cream is cooled and led to a storage tank where the fat content is analyzed and adjusted to the desired value, if necessary. The skim milk from the separator is pasteurized and cooled before being pumped to storage. It is usually destined for concentration and drying.

From the intermediate storage tanks, the cream goes to pasteurization at a temperature of 95oC or more. The high temperature is needed to destroy enzymes and micro-organisms that would impair the keeping quality of the butter.

If ripening is desired for the production of cultured butter, mixed cultures of S. cremoris, S. lactis diacetyl lactis, Leuconostocs, are used and the cream is ripened to pH 5.5 at 21 °C and then pH 4.6 at 13 °C. Most flavour development occurs between pH 5.5 - 4.6. The colder the temperature during ripening the more the flavour development relative to acid production. Ripened butter is usually not washed or salted.

In the aging tank, the cream is subjected to a program of controlled cooling designed to give the fat the required crystalline structure. The program is chosen to accord with factors such as the composition of the butterfat, expressed, for example, in terms of the iodine value which is a measure of the unsaturated fat content. The treatment can even be modified to obtain butter with good consistency despite a low iodine value, i.e. when the unsaturated proportion of the fat is low.

As a rule, aging takes 12 - 15 hours. From the aging tank, the cream is pumped to the churn or continuous buttermaker via a plate heat exchanger which brings it to the requisite temperature. In the churning process the cream is violently agitated to break down the fat globules, causing the fat to coagulate into butter grains, while the fat content of the remaining liquid, the buttermilk, decreases.

Thus the cream is split into two fractions: butter grains and buttermilk. In traditional churning, the machine stops when the grains have reached a certain size, whereupon the buttermilk is drained off. With the continuous buttermaker the draining of the buttermilk is also continuous.

After draining, the butter is worked to a continuous fat phase containing a finely dispersed water phase. It used to be common practice to wash the butter after churning to remove any residual buttermilk and milk solids but this is rarely done today.

Salt is used to improve the flavour and the shelf-life, as it acts as a preservative. If the butter is to be salted, salt (1-3%) is spread over its surface, in the case of batch production. In the continuous buttermaker, a salt slurry is added to the butter. The salt is all dissolved in the aqueous phase, so the effective salt concentration is approximately 10% in the water.

After salting, the butter must be worked vigorously to ensure even distribution of the salt. The working of the butter also influences the characteristics by which the product is judged - aroma, taste, keeping quality, appearance and colour. Working is required to obtain a homogenous blend of butter granules, water and salt. During working, fat moves from globular to free fat. Water droplets decrease in size during working and should not be visible in properly worked butter. Overworked butter will be too brittle or greasy depending on whether the fat is hard or soft. Some water may be added to standardize the moisture content. Precise control of composition is essential for maximum yield.

The finished butter is discharged into the packaging unit, and from there to cold storage.

The Background Science of Butter Churning

The Fat Globule

Milk fat is comprised mostly of triglycerides, with small amounts of mono- and diglycerides, phospholipids, glycolipids, and lipo-proteins. The triglycerides (98% of milkfat) are of diverse composition with respect to their component fatty acids, approximately 40% of which are unsaturated fat firmness varies with chain length, degree of unsaturation, and position of the fatty acids on the glycerol. Fat globules vary from 0.1 - 10 micron in diameter. The fat globule membrane is comprised of surface active materials: phospholipids and lipoproteins.

Fat globules typically aggregate in three ways:

- Flocculation,

- Coalescence,

- Partial coalescence.

Whipping and Churning

Many milk products foam easily. Skim milk foams copiously with the amount of foam being very dependent on the amount of residual fat-fat depresses foaming. The foaming agents are proteins, the amount of proteins in the foam are proportional to their contents in milk. Foaming is decreased in heat treated milk, possibly because denatured whey proteins produce a more brittle protein layer at the interface. Fats tend to spread over the air-water interface and destabilize the foam; very small amounts of fats (including phospholipids) can destabilize a foam.

During the interaction of fat globules with air bubbles the globule may also be disrupted (this is the only way that fat globules can be disrupted without considerable energy input). Disruption of the fat globule by interaction between the fat globule and air bubbles is rare except in the case of newly formed air bubbles where the air-water interfacial layer is still thin. If part of the fat globule is solid, churning will result, hence the term "flotation churning" from repeated rupturing of air bubbles and resulting coalescence of the adsorbed fat.

In spite destabilization of foams by fat, milk fat is essential for the formation of stable whipped products which depend on the interaction between fat globules, air bubbles and plasma components (especially proteins).

When cream is beaten air cells form more slowly partly because of higher viscosity and partly because the presence of fat causes immediate collapse of most of the larger bubbles. If most of the fat is liquid (high temperature) the fat globule membrane is not readily punctured and churning does not occur at cold temperature where solid fat is present, churning (clumping) of the fat globule takes place. Clumps of globules begin to associate with air bubbles so that a network of air bubbles and fat clumps and globules form entrapping all the liquid and producing a stable foam. If beating continues the fat clumps increase in size until they become too large and too few to enclose the air cells, hence air bubbles coalesce, the foam begins to leak and ultimately butter and butter milk remain.

Crystallizing of the Milkfat during Aging

Before churning, cream is subjected to a program of cooling designed to control the crystallization of the fat so that the resultant butter has the right consistency. The consistency of butter is one of its most important quality-related characteristics, both directly and indirectly, since it affects the other characteristics - chiefly taste and aroma. Consistency is a complicated concept and involves properties such as hardness, viscosity, plasticity and spreading ability.

The relative amounts of fatty acids with high melting point determine whether the fat will be hard or soft. Soft fat has a high content of low-melting fatty acids and at room temperature this fat has a large continuous fat phase with a low solid phase, i.e. crystallized, high-melting fat. On the other hand, in a hard fat, the solid phase of high-melting fat is much larger than the continuous fat phase of low-melting fatty acids.

In buttermaking, if the cream is always subjected to the same heat treatment it will be the chemical composition of the milk fat that determines the butter's consistency. A soft milk fat will make a soft and greasy butter, whereas butter from hard milk fat will be hard and stiff. If, however, the heat treatment is modified to suit the iodine value of the fat, the consistency of the butter can be optimized. For the heat treatment regulates the size of the fat crystals, and the relative amounts of solid fat and the continuous phase - the factors that determine the consistency of the butter.

Pasteurization causes the fat in the fat globules to liquefy. And when the cream is subsequently cooled a proportion of the fat will crystallize. If cooling is rapid, the crystals will be many and small; if gradual the yield will be fewer but larger crystals. The more violent the cooling process, the more will be the fat that will crystallize to form the solid phase, and the less the liquid fat that can be squeezed out of the fat globules during churning and working.

The crystals bind the liquid fat to their surface by adsorption. Since the total surface area is much greater if the crystals are many and small, more liquid fat will be adsorbed than if the crystals were larger and fewer. In the former case, churning and working will press only a small proportion of the liquid fat from the fat globules. The continuous fat phase will consequently be small and the butter firm. In the latter case, the opposite applies. A larger amount of liquid fat will be pressed out; the continuous phase will be large and the butter soft.

So by modifying the cooling program for the cream, it is possible to regulate the size of the crystals in the fat globules and in this way influence both the magnitude and the nature of the important continuous fat phase.

Tempering Treatment of Hard Fat. For optimum consistency where the iodine value is low, i.e. the butterfat is hard, as much as possible of the hardest fat must be converted to as few crystals as possible, so that little of the liquid fat is bound to the crystals. The liquid fat phase in the fat globules will thereby be maximized and much of it can be pressed out during churning and working, resulting in butter with a relatively large continuous phase of liquid fat and with the hard fat concentrated to the solid phase.

The program of treatment necessary to achieve this result comprises the following stages:

- Rapid cooling to about 8 °C and storage for about 2 hours at this temperature;

- Heating gently to 20 - 21 °C and storage at this temperature for at least 2 hours (water at 27 - 29 °C is used for heating);

- Cooling to about 16 °C.

Cooling to about 8 °C causes the formation of a large number of small crystals that bind fat from the liquid continuous phase to their surface.

When the cream is gently heated to 20 - 21 °C the bulk of the crystals melt, leaving only the hard fat crystals which, during the storage period at 20 - 21 °C, grow larger.

After 1 - 2 hours most of the hard fat has crystallized, binding little of the liquid fat. By dropping the temperature now to about 16 °C, the hardest portion of the fat will be fixed in crystal form while the rest is liquefied. During the holding period at 16 °C, fat with a melting point of 16 °C or higher will be added to the crystals. The treatment has thus caused the high-melting fat to collect in large crystals with little adsorption of the low-melting liquid fat, so that a large proportion of the butter oil can be pressed out during churning and working.

Tempering Treatment of Medium Hard Fat. With an increase in the iodine value, the heating temperature is accordingly reduced from 20-21 °C. Consequently a larger number of fat crystals will form and more liquid fat will be adsorbed than is the case with the hard fat program. For iodine values up to 39, the heating temperature can be as low as 15 °C.

Tempering Treatment of Very Soft Fat. Where the iodine value is greater than 39-40 the "summer method" of treatment is used. After pasteurization the cream is cooled to 20 °C. If the iodine value is around 39 - 40 the cream is cooled to about 8 °C, and if 41 or greater to 6 °C. It is generally held that aging temperatures below the 200 level will give a soft butter.

Butter Structure

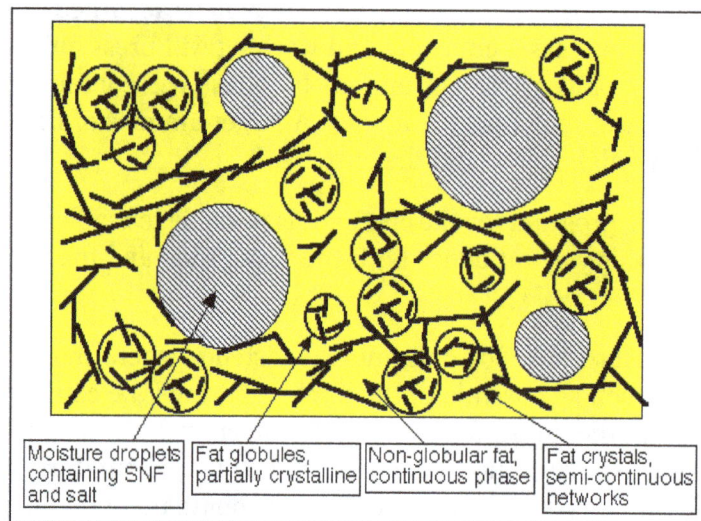

the background science of churning and the crystallization processes that the structure of butter is quite complicated. The size and extent of crystal networks both within the globules and within the non-globular phases is controlled to a large extent by milkfat's variable composition and by the aging process. The extent of globular versus non-globular fat is controlled to a large extent also by the amount of physical working applied to the butter post-churning.

Continuous Buttermaking

There are essentially four types of buttermaking processes:

- Traditional batch churning from 25- 35% mf. cream;

- Continuous flotation churning from 30-50% mf. cream;

- The concentration process whereby "plastic" cream at 82% mf. is separated from 35% mf. cream at 55 °C and then this oil-in-water emulsion cream is inverted to a water-in-oil emulsion butter with no further draining of buttermilk;

- The anhydrous milkfat process whereby water, SNF, and salt are emulsified into butter oil in a process very similar to margarine manufacture.

An optimum churning temperature must be determined for each type of process but is mainly dependent on the mean melting point and melting range of the lipids, i.e., 7-10 °C in summer and 10 - 13 °C in winter. If churning temperature is too warm or if the thermal cream aging cycle permits too much liquid fat, then a soft greasy texture results; if too cold or too much solid fat, then butter becomes too brittle.

Continuous Flotation Churns

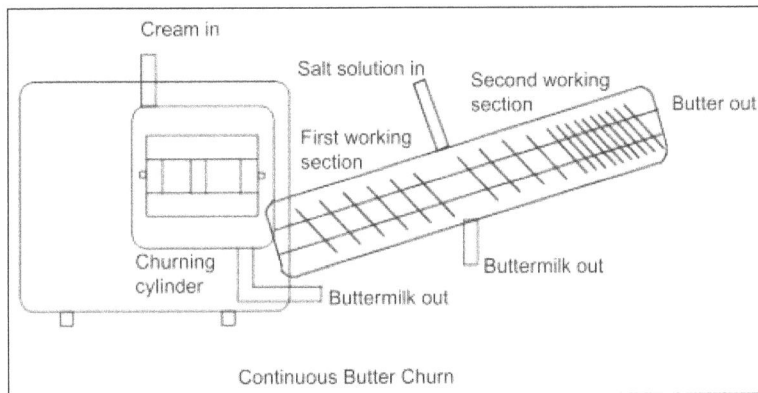

Continuous Butter Churn

The cream is first fed into a churning cylinder fitted with beaters that are driven by a variable speed motor.

Rapid conversion takes place in the cylinder and, when finished, the butter grains and buttermilk pass on to a draining section. The first washing of the butter grains sometimes takes place en route - either with water or recirculated chilled buttermilk. The working of the butter commences in the draining section by means of a screw, which also conveys it to the next stage.

On leaving the working section the butter passes through a conical channel to remove any remaining buttermilk. Immediately afterwards, the butter may be given its second washing, this time by two rows of adjustable high-pressure nozzles. The water pressure is so high that the ribbon of butter is broken down into grains and consequently any residual milk solids are effectively removed. Following this stage, salt may be added through a high-pressure injector.

The third section in the working cylinder is connected to a vacuum pump. Here it is possible to reduce the air content of the butter to the same level as conventionally churned butter.

In the final or mixing section the butter passes a series of perforated disks and star wheels. There is also an injector for final adjustment of the water content. Once regulated, the water content of the butter deviates less than +/- 0.1%, provided the characteristics of the cream remain the same.

The finished butter is discharged in a continuous ribbon from the end nozzle of the machine and then into the packaging unit.

Concentration Method

- 30% fat cream pasteurized at 90 °C.

- Degassed in a vacuum.

- Cooled to 45-70 °C.

- Separated to 82% fat ("plastic" cream).

- The concentrate, still an O/W emulsion, is cooled to 8-13 °C.

- Fat crystals forming in the tightly packed globules perforate the membranes, cause liquid fat leakage and rapid phase inversion.

- Contrast to mayonnaise, also a o/w emulsion at 82% fat but is winterized to prevent crystallization.

- Butter from this method contains all membrane material, therefore, more phospholipids.

- No butter milk produced.

- After phase inversion the butter is worked and salted.

Phase Separation

Butter from anhydrous milkfat:

- Prepare "plastic" cream (>80% fat).

- Heat with agitation to destabilize emulsion.

- Separate oil from aqueous phase: 82 to 98% butter fat.

- This butter oil is then blended with water, salt and milk solids in an emulsion pump and transferred to a scraped surface heat exchanger for cooling and to initiate crystallization.

- Further worked to develop crystal structure and texture.

- Process similar to margarine manufacture.

- Margarine has advantage of fat composition control to modify physical properties.

- Butter produced by phase separation contains few phospholipids.

Butter Yield Calculations

Technological limits to yield efficiency are defined by separation efficiency, churning efficiency, composition overrun, and package over fill.

Separation Efficiency (Es)

Represents fat transferred from milk to cream:

$$Es = 1 - fs/fm$$

where, fs = skim fat as percent w/w,

fm = milk fat as percent w/w.

Separation efficiency depends on initial milk fat content and residual fat in the skim. Assuming optimum operation of the separator, the principal determining factor of fat loss to the skim is fat globule size. Modern separators should achieve a skim fat content of 0.04 - 0.07%.

Churning Efficiency (Ec)

Represents fat transferred from cream to butter:

$$Ec = 1 - fbm/fc$$

where, fbm = buttermilk fat as percent w/w,

fc = cream fat as percent w/w.

Maximum acceptable fat loss in buttermilk is about 0.7% of churned fat corresponding to a churning efficiency of 99.3% of cream fat recovered in the butter. Churning efficiency is highest in the winter months and lowest in the summer months. Fat losses are higher in ripened butter due to a restructuring of the FGM (possibly involving crystallization of high melting triglycerides on the surface of the globules). If churning temperature is too high, churning occurs more quickly but fat loss in buttermilk increases. For continuous churns assuming 45% cream, churning efficiency should be 99.61 - 99.42%.

Composition Overrun

% Churn Overrun = (Kg butter made - Kg fat churned)/Kg fat churned x 100 %

% Composition Overrun = (100 - % fat in butter)/% fat x 100 %

Package Fill Control = (actual wt. - nominal wt.)/nominal wt. x 100%

An acceptable range for 25 kg butter blocks is 0.2 - 0.4% overfill. Overfill on 454 g prints is about 0.6%.

Other Factors Affecting Yield

- Shrinkage due to leaky butter (improperly worked).

- Shrinkage due to moisture loss; avoided by aluminum wrap.

- Loss of butter remnants on processing equipment; % loss minimal in large scale continuous processing.

Plant Overrun

Plant efficiency or plant overrun is the sum of separation, churning, composition overrun and package fill efficiencies.

- Separation Efficiency 98.85

- Churning Efficiency 99.60

- Composition overrun (% fat) 23.30

- Package overfill 0.20

These values can be used to predict the expected yield of butter per kg of milk or kg of milk fat received.

Example:

3.6% m.f. milk,

0.05% m.f. in skim,

40% m.f. in cream,

0.3% m.f. in buttermilk,

81.5% m.f. in butter,

Es = 1 - .05/3.6 = 98.6,

Ec = 1 - .3/40 = 99.25,

% Composition Overrun = (100-81.5)/81.5 = 22.7%.

If 100 kg of milk was used, 8.9 kg of cream would be produced and 4.35 kg butter would be produced from that. This is the theoretical yield based on no losses. The mass balance of fat shows that 98.3% of the fat ended up in the butter, 0.4% of the fat ended up in the buttermilk and 1.3% of the fat ended up in the skim.

The % Churn Overrun = (4.35 - 3.6)/3.6 = 20.8%.

Whipped Butter

Whipped butter is typically used in foodservice situations. The main advantage of whipped butter is increased spreadability even at refrigeration temperatures, thus providing great advantage for the restaurant industry. The volume increase is usually 25 - 30%. Whipping is achieved by injecting an inert gas (nitrogen) into the butter after churning. In the phase separation process, whipping can be achieved by injecting nitrogen in the crystallizer as is done in the production of whipped margarine.

Anhydrous Milkfat (Butter Oil)

Anhydrous milk fat, butter oil, can be manufactured from either butter or from cream. For the manufacture from butter, non-salted butter from sweet cream is normally used, and the process works better if the butter is at least a few weeks old. Melted butter is passed through a centrifuge, to concentrate the fat to 99.5% of greater. This oil is heated again to 90-95 °C and vacuum cooled before packaging.

The processes for the production of anhydrous fat, using cream as the raw material, are based on the emulsion splitting principle. In brief, the processes consist of the cream first being concentrated to 75% fat or greater, in two stages. In both of these stages, the fat is concentrated in a hermetic solids-ejecting separator. The fat globules are then broken down mechanically, so that phase inversion occurs and the fat is liberated. This forms a continuous fat phase containing dispersed water droplets, which can be separated from the fat phase by centrifugation. This is similar to the concentration method for buttermaking, with the addition of the mechanical rupture of the emulsion and additional separator for removal of the residual water phase.

One of the key machines in the system is the mechanical device for phase inversion. This can be in the form of a centrifugal separator equipped with a serrated disc. The disc breaks down the emulsion, so that the liquid leaving the machine is a continuous oil phase, with dispersed water droplets and buttermilk. Larger equipment could be equipped with a motor-driven serrated disc or with a homogenizer. After phase inversion, the fat is concentrated to 99.5% or greater in a hermetic separator.

Fractionation of Anhydrous Milk Fat

Milk fat is a complicated mixture of triglycerides that contain numerous fatty acids of varying carbon chain lengths and degrees of saturation. The proportions of the various fatty acids present will also vary depending on the conditions surrounding the production of milk.

One method of milkfat fraction is by thermal treatment. The mixture can be separated into fractions on the basis of their melting point. The technique consists of melting the entire quantity of fat and then cooling it down to a predetermined temperature. The triglycerides with the higher melting point will then crystallize and settle out.

In the modern thermal fractionation method, sedimentation by gravity is replaced by centrifugal separation. Since a modern separator generates a force that is thousands of times greater than the force of gravity and since the sedimentation distances are very short, the process is incomparably faster. The crystallizing stage can also be accelerated, since the crystals need not be large if centrifugal separation is employed.

Fractionation of milkfat can also be accomplished by supercritical fluid extraction techniques.

CUSTARD

Custard is a variety of culinary preparations based on milk or cream cooked with egg yolk to thicken it, and sometimes also flour, corn starch, or gelatin. Depending on the recipe, custard may vary in consistency from a thin pouring sauce (crème anglaise) to the thick pastry cream (crème pâtissière) used to fill éclairs. The most common custards are used in desserts or dessert sauces and typically include sugar and vanilla, however savory custards are also found, e.g. in quiche.

Custard is usually cooked in a double boiler (bain-marie), or heated very gently in a saucepan on a stove, though custard can also be steamed, baked in the oven with or without a water bath, or even cooked in a pressure cooker. Custard preparation is a delicate operation, because a temperature increase of 3–6 °C (5–10 °F) leads to overcooking and curdling. Generally, a fully cooked custard should not exceed 80 °C (~175 °F); it begins setting at 70 °C (~160 °F). A water bath slows heat transfer and makes it easier to remove the custard from the oven before it curdles. Adding a small amount of cornflour to the egg-sugar mixture stabilises the resulting custard, allowing it to be cooked in a single pan as well as in a double-boiler. A sous-vide water bath may be used to precisely control temperature.

Pastry cream.   A bowl of custard.

Mixtures of milk and eggs thickened by heat have long been part of European cuisine, since at least Ancient Rome. Custards baked in pastry (custard tarts) were very popular in the Middle Ages.

Examples include Crustardes of flessh and Crustade, in the 14th century English collection The Forme of Cury. These recipes include solid ingredients such as meat, fish, and fruit bound by the custard. Stirred custards cooked in pots are also found under the names Creme Boylede and Creme boiled.

In modern times, the name 'custard' is sometimes applied to starch-thickened preparations like blancmange and Bird's Custard powder.

Custard Variations

While custard may refer to a wide variety of thickened dishes, technically (and in French cookery) the word "custard".

When starch is added, the result is called pastry cream or confectioners' custard, made with a combination of milk or cream, egg yolks, fine sugar, flour or some other starch, and usually a flavoring such as vanilla, chocolate, or lemon. Crème pâtissière is a key ingredient in many French desserts including mille-feuille (or Napoleons) and filled tarts. It is also used in Italian pastry and sometimes in Boston cream pie. The thickening of the custard is caused by the combination of egg and starch. Corn flour or flour thicken at 100 °C (212 °F) and as such many recipes instruct the pastry cream to be boiled. In a traditional custard such as a crème anglaise, where egg is used alone as a thickener, boiling results in the over cooking and subsequent 'curdling' of the custard; however, in a pastry cream, starch prevents this. Once cooled, the amount of starch in pastry cream 'sets' the cream and requires it to be beaten or whipped before use.

A formal custard preparation, garnished with raspberries.

When gelatin is added, it is known as crème anglaise collée. When gelatin is added and whipped cream is folded in, and it sets in a mold, it is bavarois. When starch is used alone as a thickener (without eggs), the result is a blancmange. In the United Kingdom, custard has various traditional recipes some thickened principally with cornflour (cornstarch) rather than the egg component, others involving regular flour.

After the custard has thickened, it may be mixed with other ingredients: mixed with stiffly beaten egg whites and gelatin, it is chiboust cream; mixed with whipped cream, Beating in softened butter produces German buttercream or crème mousseline.

Layers of a trifle showing the custard in
between cake, fruit & whipped cream.

A quiche is a savoury custard tart. Some kinds of timbale or vegetable loaf are made of a custard base mixed with chopped savoury ingredients. Custard royale is a thick custard cut into decorative shapes and used to garnish soup, stew or broth. In German it is known as Eierstich and is used as a garnish in German Wedding Soup (Hochzeitssuppe). Chawanmushi is a Japanese savoury custard, steamed and served in a small bowl or on a saucer. Chinese steamed egg is a similar but larger savoury egg dish. Bougatsa is a Greek breakfast pastry whose sweet version consists of semolina custard filling between layers of phyllo.

Custard may also be used as a top layer in gratins, such as the South African bobotie and many Balkan versions of moussaka.

Physical-chemical Properties

Cooked (set) custard is a weak gel, viscous and thixotropic; while it does become easier to stir the more it is manipulated, it does not, unlike many other thixotropic liquids, recover its lost viscosity over time. On the other hand, a suspension of uncooked imitation custard powder (starch) in water, with the proper proportions, has the opposite rheological property: it is negative thixotropic, or dilatant, allowing the demonstration of "walking on custard".

Chemistry

Eggs contain the proteins necessary for the gel structure to form, and emulsifiers to maintain the structure. Egg yolk also contains enzymes like amylase, which can break down added starch. This enzyme activity contributes to the overall thinning of custard in the mouth. Egg yolk lecithin also helps to maintain the milk-egg interface. The proteins in egg whites set at 60-80 °C (140-180 °F).

Starch is sometimes added to custard to prevent premature curdling. The starch acts as a heat buffer in the mixture: as they hydrate, they absorb heat and help maintain a constant rate of heat transfer. Starches also make for a smoother texture and thicker mouthfeel.

If the mixture pH is 9 or higher, the gel is too hard; if it is below 5, the gel structure has difficulty forming because of protonation prevents the formation of covalent bonds.

ICE CREAM

Ice cream (derived from earlier iced cream or cream ice) is a sweetened frozen food typically eaten as a snack or dessert. It may be made from dairy milk or cream and is flavored with a sweetener, either sugar or an alternative, and any spice, such as cocoa or vanilla. Colourings are usually added, in addition to stabilizers. The mixture is stirred to incorporate air spaces and cooled below the freezing point of water to prevent detectable ice crystals from forming. The result is a smooth, semi-solid foam that is solid at very low temperatures (below 2 °C or 35 °F). It becomes more malleable as its temperature increases.

The meaning of the name "ice cream" varies from one country to another. Terms such as "frozen custard," "frozen yogurt," "sorbet," "gelato," and others are used to distinguish different varieties and styles. In some countries, such as the United States, "ice cream" applies only to a specific variety, and most governments regulate the commercial use of the various terms according to the relative quantities of the main ingredients, notably the amount of cream. Products that do not meet the criteria to be called ice cream are sometimes labelled "frozen dessert" instead. In other countries, such as Italy and Argentina, one word is used for all variants. Analogues made from dairy alternatives, such as goat's or sheep's milk, or milk substitutes (e.g., soy, cashew, coconut, almond milk or tofu), are available for those who are lactose intolerant, allergic to dairy protein, or vegan.

Ice cream may be served in dishes, for eating with a spoon, or licked from edible cones. Ice cream may be served with other desserts, such as apple pie, or as an ingredient in ice cream floats, sundaes, milkshakes, ice cream cakes and even baked items, such as Baked Alaska.

Composition

Ice cream is a colloidal emulsion having dispersed phase as fat globules. It is an emulsion which is in the end made into foam by incorporating air cells which is frozen to form dispersed ice cells. In the composition of ice cream ice crystals are of most importance as they give a desirable mouth feel. Ice cream is composed of water, ice, milk fat, milk protein, sugar and air. Water and fat have highest proportions by weight creating an emulsion. The triacylglycerols in fat are non polar and will adhere to itself by van der Waals interactions. Water is polar thus, emulsifiers are needed for dispersion of fat. Also ice cream has a colloidal phase of foam which helps in light texture. Milk proteins such as casein and whey protein present in ice cream are amphiphilic, can adsorb water and form micelles which will contribute to consistency. Sucrose which is disaccharide is usually used as a sweetening agent. Lactose which is sugar present in milk will cause freezing point depression. Thus, on freezing some water will be unfrozen and will not give hard texture.

Production

Ice cream maker Boku Europa.

Before the development of modern refrigeration, ice cream was a luxury reserved for special occasions. Making it was quite laborious; ice was cut from lakes and ponds during the winter and stored in holes in the ground, or in wood-frame or brick ice houses, insulated by straw. Many farmers and plantation owners, including U.S. Presidents George Washington and Thomas Jefferson, cut and stored ice in the winter for use in the summer. Frederic Tudor of Boston turned ice harvesting and shipping into a big business, cutting ice in New England and shipping it around the world.

Ice cream was made by hand in a large bowl placed inside a tub filled with ice and salt. This is called the pot-freezer method. French confectioners refined the pot-freezer method, making ice cream in a sorbetière (a covered pail with a handle attached to the lid). In the pot-freezer method, the temperature of the ingredients is reduced by the mixture of crushed ice and salt. The salt water is cooled by the ice, and the action of the salt on the ice causes it to (partially) melt, absorbing latent heat and bringing the mixture below the freezing point of pure water. The immersed container can also make better thermal contact with the salty water and ice mixture than it could with ice alone.

The hand-cranked churn, which also uses ice and salt for cooling, replaced the pot-freezer method. The exact origin of the hand-cranked freezer is unknown, but the first U.S. patent for one was #3254 issued to Nancy Johnson on 9 September 1843. The hand-cranked churn produced smoother ice cream than the pot freezer and did it quicker. Many inventors patented improvements on Johnson's design.

In Europe and early America, ice cream was made and sold by small businesses, mostly confectioners and caterers. Jacob Fussell of Baltimore, Maryland was the first to manufacture ice cream on a large scale. Fussell bought fresh dairy products from farmers in York County, Pennsylvania, and sold them in Baltimore. An unstable demand for his dairy products often left him with a surplus of cream, which he made into ice cream. He built his first ice cream factory in Seven Valleys, Pennsylvania, in 1851. Two years later, he moved his factory to Baltimore. Later, he opened factories in several other cities and taught the business to others, who operated their own plants. Mass production reduced the cost of ice cream and added to its popularity.

The development of industrial refrigeration by German engineer, Carl von Linde during the 1870s eliminated the need to cut and store natural ice, and, when the continuous-process freezer was perfected in 1926, commercial mass production of ice cream and the birth of the modern ice cream industry was underway.

In modern times, a common method for producing ice cream at home is to use an ice cream maker, an electrical device that churns the ice cream mixture while cooled inside a household freezer. Some more expensive models have an inbuilt freezing element. A newer method is to add liquid nitrogen to the mixture while stirring it using a spoon or spatula for a few seconds; a similar technique, advocated by Heston Blumenthal as ideal for home cooks, is to add dry ice to the mixture while stirring for a few minutes. Some ice cream recipes call for making a custard, folding in whipped cream, and immediately freezing the mixture. Another method is to use a pre-frozen solution of salt and water, which gradually melts as the ice cream freezes.

Borden's Eagle Brand sweetened condensed milk circulated a recipe for making ice cream at home. It may be made in an ice cube tray with condensed milk, cream, and various simple flavorings. It can be ready to serve after as little as four hours of freezing. Fresh or frozen fruit, nuts, chocolate, and other ingredients may be added as well.

An unusual method of making ice-cream was done during World War II by American fighter pilots based in the South Pacific. They attached pairs of 5-US-gallon (19 l) cans to their aircraft. The cans were fitted with a small propeller, this was spun by the slipstream and drove a stirrer, which agitated the mixture while the intense cold of high altitude froze it. B-17 crews in Europe did something similar on their bombing runs as did others.

Retail Sales

Ice cream can be mass-produced and thus is widely available in developed parts of the world. Ice cream can be purchased in large cartons (vats and squrounds) from supermarkets and grocery stores, in smaller quantities from ice cream shops, convenience stores, and milk bars, and in individual servings from small carts or vans at public events. In 2015, the US produced nearly 900 million gallons of ice cream.

A bicycle-based ice cream street vendor in Indonesia.

Specialty Job

Today, jobs specialize in the selling of ice cream. The title of a person who works in this speciality is often called an 'ice cream man', however women also specialize in the selling of ice cream. People in this line of work often sell ice cream on beaches. On beaches, ice cream is either sold by a person who carries a box full of ice cream and is called over by people who want the purchase ice cream, or by a person who drives up to the top of the beach and rings a bell. In the second method, people go up to the top of the beach and purchase ice cream straight from the ice cream seller, who is often in an ice cream van. In Turkey and Australia, ice cream is sometimes sold to beach-goers from small powerboats equipped with chest freezers.

Ice cream van vendor delivery.

Some ice cream distributors sell ice cream products from traveling refrigerated vans or carts (commonly referred to in the US as "ice cream trucks"), sometimes equipped with speakers playing children's music or folk melodies (such as "Turkey in the Straw"). The driver of an ice cream van drives throughout neighborhoods and stops every so often, usually every block. The seller on the ice cream van sells the ice cream through a large window; this window is also where the customer asks for ice cream and pays. Ice cream vans in the United Kingdom make a music box noise rather than actual music.

Ingredients and Standard Quality Definitions

Black sesame soft ice cream, Japan.

Many countries have regulations controlling what can be described as ice cream.

In the U.S., the FDA rules state that to be described as "ice cream", a product must have the following composition:

- Greater than 10% milkfat.

- 6 to 10% milk non-fat milk solids: this component, also known as the milk solids-not-fat or serum solids, contains the proteins (caseins and whey proteins) and carbohydrates (lactose) found in milk.

It generally also has:

- 12 to 16% sweeteners: usually a combination of sucrose and glucose-based corn syrup sweeteners.

- 0.2 to 0.5% stabilisers and emulsifiers.

- 55% to 64% water, which comes from the milk or other ingredients.

These compositions are percentage by weight. Since ice cream can contain as much as half air by volume, these numbers may be reduced by as much as half if cited by volume. In terms of dietary considerations, the percentages by weight are more relevant. Even the low-fat products have high caloric content: Ben and Jerry's No-Fat Vanilla Fudge contains 150 calories (630 kJ) per half-cup due to its high sugar content.

According to Canadian Food and Drugs Act and Regulations, ice cream in Canada is divided into Ice cream mix and Ice cream. Each have a different set of regulations.

- Ice cream must be at least 10 percent milk fat, and must contain at least 180 grams of solids per liter. When cocoa, chocolate syrup, fruit, nuts, or confections are added, the percentage of milk fat can be 8 percent.

- The Ice cream mix is defined as the pasteurized mix of cream, milk and other milk products that are not yet frozen. It may contain eggs, artificial or non-artificial flavours, cocoa or chocolate syrup, a food color, an agent that adjusts the pH level in the mix, salt, a stabilizing agent that doesn't exceed 0.5% of the ice cream mix, a sequestering agent which preserves the food colour, edible casein that doesn't exceed 1% of the mix, propylene glycol mono fatty acids in an amount that will not exceed 0.35% of the ice cream mix and sorbitan tristearate in an amount that will not exceed 0.035% of the mix. The ice cream mix may not include less than 36% solid components.

Physical Properties

Ice cream sandwich.

Ice cream is considered as a colloidal system. It is composed by ice cream crystals and aggregates, air that does not mix with the ice cream by forming small bubbles in the bulk and partially coalesced fat globules. This dispersed phase made from all the small particles is surrounded by an unfrozen continuous phase composed by sugars, proteins, salts, polysaccharides and water. Their interactions determine the properties of ice cream, whether soft and whippy or hard.

Ostwald Ripening

Chocolate-glazed ice cream bar.

Ostwald ripening is the explanation for the growth of large crystals at the expense of small ones in the dispersion phase. This process is also called migratory recrystallization. It involves the formation of sharp crystals. Theories about Ostwald recrystallization admit that after a period of time, the recrystallization process can be described by the following equation:

$$r = r(0) + Rt \exp(1/n)$$

Where $r(0)$ is the initial size, n the order of recrystallization, t a time constant for recrystallization that depends on the rate R (in units of size/ time).

To make ice cream smooth, recrystallization must occur as slowly as possible, because small crystals create smoothness, meaning that r must decrease.

Around the World

Around the world, different cultures have developed unique versions of ice cream, suiting the product to local tastes and preferences.

Italian ice cream, gelato in Rome, Italy.

The most traditional Argentine helado (ice cream) is very similar to Italian gelato, in large part due to the historical influence of Italian immigrants on Argentinian customs.

Per capita, Australians and New Zealanders are among the leading ice cream consumers in the world, eating 18 litres and 20 litres each per year respectively, behind the United States where people eat 23 litres each per year.

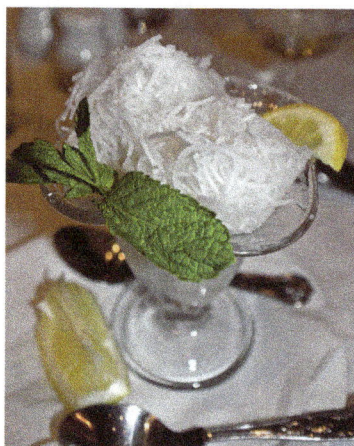

In Iran, the most popular ice cream-like treat is faludeh (also called paloodeh, paludeh or fālūdhaj), which contains vermicelli noodles, sugar syrup and rose water. It's often served with lime juice and sometimes ground pistachios.

In China, besides the popular flavours such as vanilla, chocolate, coffee, mango and strawberry, many Chinese ice-cream manufacturers have also introduced other traditional Chinese flavours such as black sesame and red bean.

In 1651, Italian Francesco dei Coltelli opened an ice cream café in Paris, and the product became so popular that during the next 50 years another 250 cafés opened in Paris.

In Greece, ice cream in its modern form, or pagotó, was introduced in the beginning of the 20th century.

India is one of the largest producers of ice cream in the world, but most of its ice cream is consumed domestically.

In Indonesia, a type of traditional ice cream called *es puter* or "stirred ice cream" is made from coconut milk, pandanus leaves, sugar and flavors that include avocado, jackfruit, durian, palm sugar, chocolate, red bean, and mung bean.

In Iran, fālūde or pālūde is a Persian sorbet made of thin vermicelli noodles, frozen with sugar syrup and rose water. The dessert is often served with lime juice and sometimes ground pistachios.

Italian ice cream, or *gelato* as it is known, is a traditional and a popular dessert in Italy. Much of the production is still hand-made and flavoured by each individual shop in "produzione propria" gelaterias. Gelato is made from whole milk, sugar, sometimes eggs, and natural flavourings. Gelato typically contains 7–8% fat, less than ice cream's minimum of 10%.

Sorbetes is a Philippine version for common ice cream usually peddled from carts by peddlers who roam streets in the Philippines. Despite the similarities between the name sorbetes and sorbet, sorbetes is not a type of sorbet.

In Spain, ice cream is often in the style of Italian gelato. Spanish gelato can be found in many cafes or specialty ice cream stores. While many traditional flavours are sold, cafes may also sell unique flavours like nata, crema catalana, or tiramisu.

Dondurma is the name given to ice cream in Turkey. Dondurma typically includes milk, sugar, salep, and mastic.

In the United Kingdom, 14 million adults buy ice cream as a treat, in a market worth £1.3 billion.

In the United States, ice cream made with just cream, sugar, and a flavouring (usually fruit) is sometimes referred to as "Philadelphia style" ice cream. Ice cream that uses eggs to make a custard is sometimes called "French ice cream". American federal labeling standards require ice cream to contain a minimum of 10% milk fat. Americans consume about 23 liters of ice cream per person per year — the most in the world.

Other Frozen Desserts

Bananas Foster flaming ice cream.

Kulfi from India with strawberry sauce.

The following is a partial list of ice cream-like frozen desserts and snacks:

- Ais kacang: A dessert in Malaysia and Singapore made from shaved ice, syrup, and boiled red bean and topped with evaporated milk. Sometimes, other small ingredients like raspberries and durians are added in, too.

- Booza: An elastic, sticky, high level melt resistant ice cream.

- Dondurma: Turkish ice cream, made of salep and mastic resin.

- Frozen custard: At least 10% milk fat and at least 1.4% egg yolk and much less air beaten into it, similar to Gelato, fairly rare. Known in Italy as Semifreddo.

- Frozen yogurt: Made with yogurt instead of milk or cream, it has a tart flavour and lower fat content.

- Gelato: An Italian frozen dessert having a lower milk fat content than ice cream.

- Halo-halo: A popular Filipino dessert that is a mixture of shaved ice and milk to which are added various boiled sweet beans and fruits, and served cold in a tall glass or bowl.

- Ice cream sandwich: Two (usually) soft biscuits, cookies or cake sandwiching a bar of ice cream.

- Ice milk: Less than 10% milk fat and lower sweetening content, once marketed as "ice milk" but now sold as low-fat ice cream in the United States.

- Ice pop or ice lolly: Frozen fruit puree, fruit juice, or flavoured sugar water on a stick or in a flexible plastic sleeve.

- Kulfi: Believed to have originated in the Indian subcontinent during Mughal India.

- Maple toffee: Also known as maple taffy. A popular springtime treat in maple-growing areas is maple toffee, where maple syrup boiled to a concentrated state is poured over fresh snow congealing in a toffee-like mass, and then eaten from a wooden stick used to pick it up from the snow.

- Mellorine: Non-dairy, with vegetable fat substituted for milk fat.

- Parevine: Kosher non-dairy frozen dessert established in 1969 in New York.

- Patbingsu: A popular Korean shaved ice dessert commonly served with sweet toppings such as fruit, red bean, or sweetened condensed milk.

- Pop up ice cream.

- Sherbet: 1–2% milk fat and sweeter than ice cream.

- Sorbet: Fruit puree with no dairy products.

- Snow cones, made from balls of crushed ice topped with flavoured syrup served in a paper cone, are consumed in many parts of the world. The most common places to find snow cones in the United States are at amusement parks.

Cryogenics

Dippin' Dots Rainbow Ice Cream.

In 2006, some commercial ice cream makers began to use liquid nitrogen in the primary freezing of ice cream, thus eliminating the need for a conventional ice cream freezer. The preparation results in a column of white condensed water vapour cloud. The ice cream, dangerous to eat while still "steaming" with liquid nitrogen, is allowed to rest until the liquid nitrogen is completely vapourised. Sometimes ice cream is frozen to the sides of the container, and must be allowed to thaw. Good results can also be achieved with the more readily available dry ice, and authors such as Heston Blumenthal have published recipes to produce ice cream and sorbet using a simple blender.

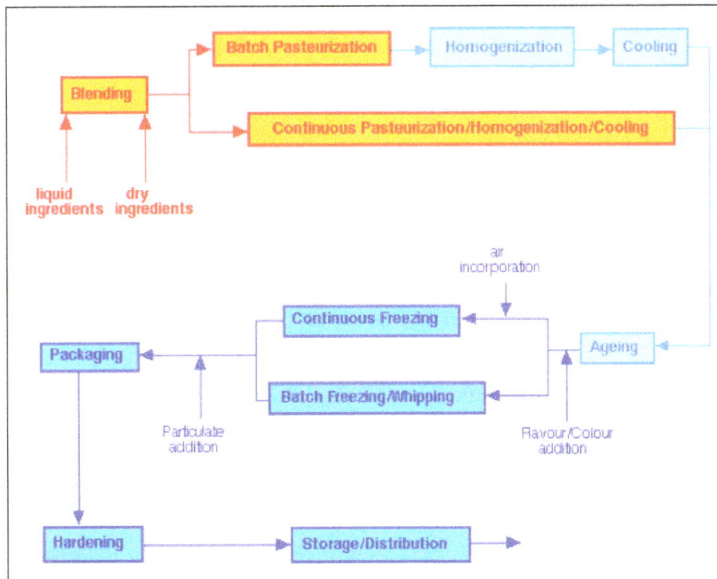

Another vendor, Creamistry, creates ice cream from liquid ingredients as customers watch. It has a softer texture than regular ice cream, because ice crystals have less time to form.

The basic steps in the manufacturing of ice cream are generally as follows:

- Blending of the mix ingredients.

- Pasteurization.

- Homogenization.

- Aging the mix.

- Freezing.

- Packaging.

- Hardening.

Process flow diagram for ice cream manufacture: the red section represents the operations involving raw, unpasteurized mix, the pale blue section represents the operations involving pasteurized mix, and the dark blue section represents the operations involving frozen ice cream.

Blending

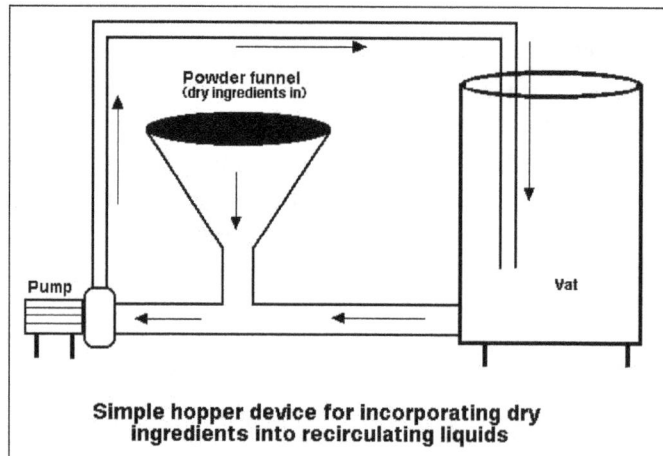

Simple hopper device for incorporating dry ingredients into recirculating liquids

First the ingredients are selected based on the desired formulation and the calculation of the recipe from the formulation and the ingredients chosen, then the ingredients are weighed and blended together to produce what is known as the "ice cream mix". Blending requires rapid agitation to incorporate powders, and often high speed blenders are used.

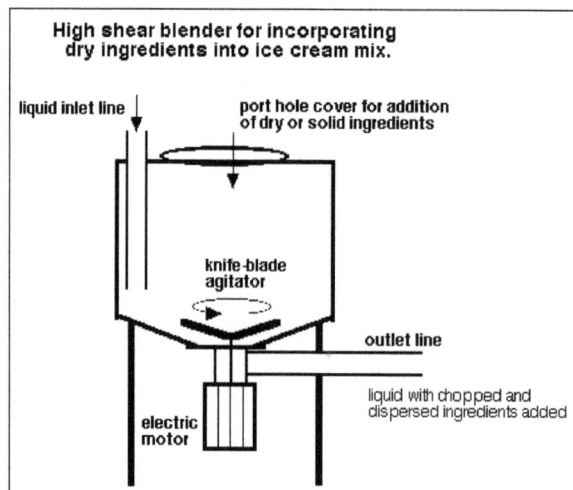

Pasteurization of Mix

The mix is then pasteurized. Pasteurization is the biological control point in the system, designed for the destruction of pathogenic bacteria. In addition to this very important function, pasteurization also reduces the number of spoilage organisms such as psychrotrophs, and helps to hydrate some of the components (proteins, stabilizers).

- Pasteurization (Ontario regulations): 69 °C/30 min. 80 °C/25s.

- Both batch pasteurizers and continuous (HTST) methods are used.

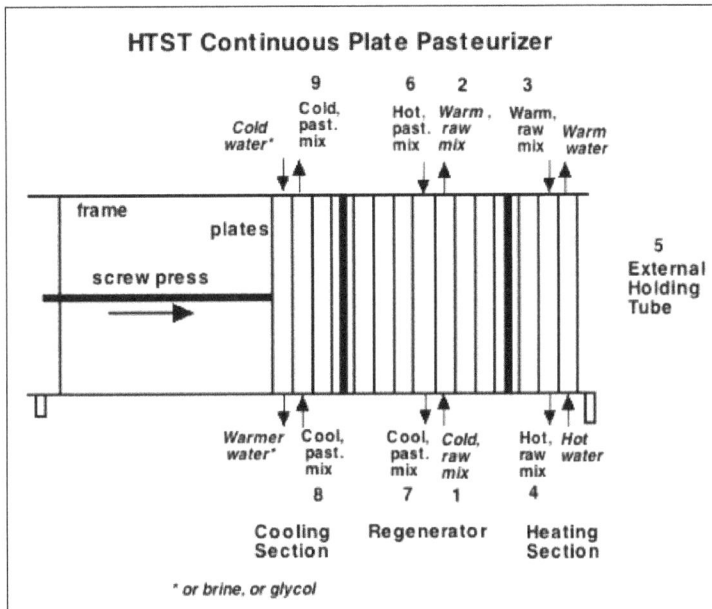

Batch pasteurizers lead to more whey protein denaturation, which some people feel gives a better body to the ice cream. In a batch pasteurization system, blending of the proper ingredient amounts is done in large jacketed vats equipped with some means of heating, usually steam or hot water. The product is then heated in the vat to at least 69 °C (155 °F) and held for 30 diagram of an HTST continuous plate pasteurizerminutes to satisfy legal requirements for pasteurization, necessary for the destruction of pathogenic bacteria. Various time temperature combinations can be used. The heat treatment must be severe enough to ensure destruction of pathogens and to reduce the bacterial count to a maximum of 100,000 per gram. Following pasteurization, the mix is homogenized by means of high pressures and then is passed across some type of heat exchanger (plate or double or triple tube) for the purpose of cooling the mix to refrigerated temperatures (4 °C). Batch tanks are usually operated in tandem so that one is holding while the other is being prepared. Automatic timers and valves ensure the proper holding time has been met.

Continuous pasteurization is usually performed in a high temperature short time (HTST) heat exchanger following blending of ingredients in a large, insulated feed tank. Some preheating, to 30 to 40 °C, is necessary for solubilization of the components. The HTST system is equipped with a heating section, a cooling section, and a regeneration section. Cooling sections of ice cream mix HTST presses are usually larger than milk HTST presses. Due to the preheating of the mix, regeneration is lost and mix entering the cooling section is still quite warm.

Homogenization of Mix

The mix is also homogenized, which forms the fat emulsion by breaking down or reducing the size of the fat globules found in milk or cream to less than 1 μ m. Two stage homogenization is usually preferred for ice cream mix. Clumping or clustering of the fat is reduced thereby producing a thinner, more rapidly whipped mix. Melt-down is also improved. Homogenization provides the following functions in ice cream manufacture:

- Reduces size of fat globules.

- Increases surface area.

- Forms membrane.

- Makes possible the use of butter, frozen cream, etc.

By helping to form the fat structure, it also has the following indirect effects:

- Makes a smoother ice cream.

- Gives a greater apparent richness and palatability.

- Better air stability.

- Increases resistance to melting.

Homogenization of the mix should take place at the pasteurizing temperature. The high temperature produces more efficient breaking up of the fat globules at any given pressure and also reduces fat clumping and the tendency to thick, heavy bodied mixes. No one pressure can be recommended that will give satisfactory results under all conditions. The higher the fat and total solids in the mix, the lower the pressure should be. If a two stage homogenizer is used, a pressure of 2000 - 2500 psi on the first stage and 500 - 1000 psi on the second stage should be satisfactory under most conditions. Two stage homogenization is usually preferred for ice cream mix. Clumping or clustering of the fat is reduced thereby producing a thinner, more rapidly whipped mix. Melt-down is also improved.

Ageing of Mix

The mix is then aged for at least four hours and usually overnight. This allows time for the fat to cool down and crystallize, and for the proteins and polysaccharides to fully hydrate. Aging provides the following functions:

- Improves whipping qualities of mix and body and texture of ice cream.

It does so by:

- Providing time for fat crystallization, so the fat can partially coalesce.

- Allowing time for full protein and stabilizer hydration and a resulting slight viscosity increase.

- Allowing time for membrane rearrangement and protein/emulsifier interaction, as emulsifiers displace proteins from the fat globule surface, which allows for a reduction in stabilization of the fat globules and enhanced partial coalescence.

Aging is performed in insulated or refrigerated storage tanks, silos, etc. Mix temperature should be maintained as low as possible without freezing, at or below 5 °C. An aging time of overnight is likely to give best results under average plant conditions. A green or unaged mix is usually quickly detected at the freezer.

Freezing/Whipping of Ice Cream

Following mix processing, the mix is drawn into a flavour tank where any liquid flavours, fruit purees, or colours are added. The mix then enters the dynamic freezing process which both freezes a portion of the water and whips air into the frozen mix. The "barrel" freezer is a scraped-surface, tubular heat exchanger, which is jacketed with a boiling refrigerant such as ammonia or freon. Mix is pumped through this freezer and is drawn off the other end in a matter of 30 seconds, (or 10 to 15 minutes in the case of batch freezers) with about 50% of its water frozen. There are rotating blades inside the barrel that keep the ice scraped off the surface of the freezer and also dashers inside the machine which help to whip the mix and incorporate air.

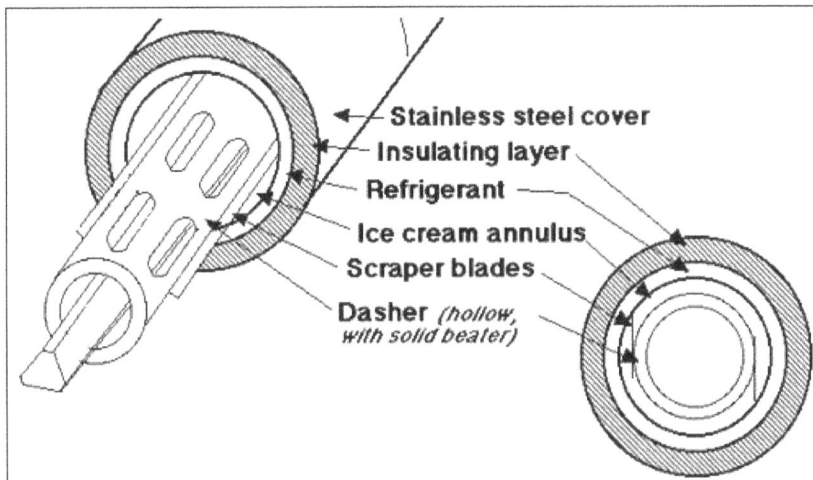

Ice cream contains a considerable quantity of air, up to half of its volume. This gives the product its characteristic lightness. Without air, ice cream would be similar to a frozen ice cube. The air content is termed its overrun, which can be calculated mathematically.

As the ice cream is drawn with about half of its water frozen, particulate matter such as fruits, nuts, candy, cookies, or whatever you like, is added to the semi-frozen slurry which has a consistency similar to soft-serve ice cream. In fact, almost the only thing which differentiates hard frozen ice cream from soft-serve, is the fact that soft serve is drawn into cones at this point in the process rather than into packages for subsequent hardening.

Hardening

After the particulates have been added, the ice cream is packaged and is placed into a blast freezer at -30° to -40 °C where most of the remainder of the water is frozen. Below about -25 °C, ice cream is stable for indefinite periods without danger of ice crystal growth; however, above this temperature, ice crystal growth is possible and the rate of crystal growth is dependant upon the temperature of storage. This limits the shelf life of ice cream.

A primer on the theoretical aspects of freezing will help you to fully understand the freezing and recrystallization process.

Hardening invloves static (still, quiescent) freezing of the packaged products in blast freezers. Freezing rate must still be rapid, so freezing techniques involve low temperature (-40 °C) with either enhanced convection (freezing tunnels with forced air fans) or enhanced conduction (plate freezers).

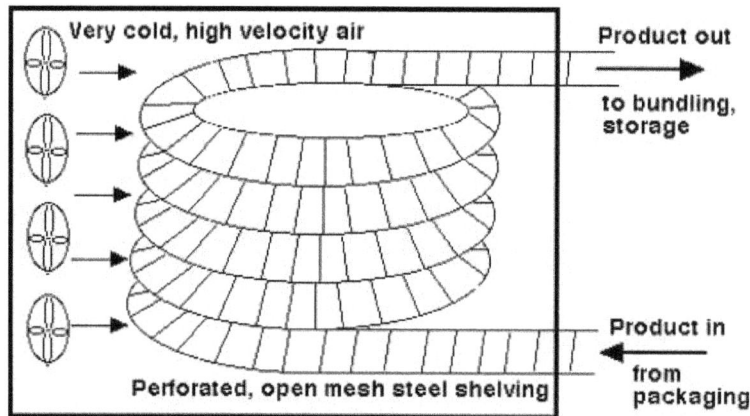

The rate of heat transfer in a freezing process is affected by the temperature difference, the surface area exposed and the heat transfer coefficient (Q=U A dT). Thus, the factors affecting hardening are those affecting this rate of heat transfer:

- Temperature of blast freezer: The colder the temperature, the faster the hardening, the smoother the product.

- Rapid circulation of air: Increases convective heat transfer.

- Temperature of ice cream when placed in the hardening freezer: The colder the ice cream at draw, the faster the hardening; must get through packaging operations fast.

- Size of container: Exposure of maximum surface area to cold air, especially important to consider shrink wrapped bundles, they become a much larger mass to freeze. Bundling should be done after hardening.

- Composition of ice cream: Related to freezing point depression and the temperature required to ensure a significantly high ice phase volume.

- Method of stacking containers or bundles to allow air circulation. Circulation should not be impeded there should be no 'dead air' spaces (e.g., round vs. square packages).

- Care of evaporator: Freedom from frost acts as insulator.

- Package type should not impede heat transfer e.g., styrofoam liner or corrugated cardboard may protect against heat shock after hardening, but reduces heat transfer during freezing so not feasible.

Ice cream from the dynamic freezing process (continuous freezer) can also be transformed into an array of novely/impulse products through a variety of filling and forming machines.

CHEESE

Red Hawk cheese.

A platter with cheese and garnishes.

Cheese is a dairy product derived from milk that is produced in a wide range of flavors, textures, and forms by coagulation of the milk protein casein. It comprises proteins and fat from milk, usually the milk of cows, buffalo, goats, or sheep. During production, the milk is usually acidified, and adding the enzyme rennet causes coagulation. The solids are separated and pressed into final form. Some cheeses have molds on the rind, the outer layer, or throughout. Most cheeses melt at cooking temperature.

Over a thousand types of cheese from various countries are produced. Their styles, textures and flavors depend on the origin of the milk (including the animal's diet), whether they have been pasteurized, the butterfat content, the bacteria and mold, the processing, and aging. Herbs, spices, or wood smoke may be used as flavoring agents. The yellow to red color of many cheeses, such as Red Leicester, is produced by adding annatto. Other ingredients may be added to some cheeses, such as black pepper, garlic, chives or cranberries.

Cheeses in art: Still Life with Cheeses,
Almonds and Pretzels, Clara Peeters.

For a few cheeses, the milk is curdled by adding acids such as vinegar or lemon juice. Most cheeses are acidified to a lesser degree by bacteria, which turn milk sugars into lactic acid, then the addition of rennet completes the curdling. Vegetarian alternatives to rennet are available; most are produced by fermentation of the fungus Mucor miehei, but others have been extracted from various species of the Cynara thistle family. Cheesemakers near a dairy region may benefit from fresher, lower-priced milk, and lower shipping costs.

Cheese is valued for its portability, long life, and high content of fat, protein, calcium, and phosphorus. Cheese is more compact and has a longer shelf life than milk, although how long a cheese will keep depends on the type of cheese. Hard cheeses, such as Parmesan, last longer than soft cheeses, such as Brie or goat's milk cheese. The long storage life of some cheeses, especially when encased in a protective rind, allows selling when markets are favorable. Vacuum packaging of block-shaped cheeses and gas-flushing of plastic bags with mixtures of carbon dioxide and nitrogen are used for storage and mass distribution of cheeses in the 21st century.

A specialist seller of cheese is sometimes known as a cheesemonger. Becoming an expert in this field requires some formal education and years of tasting and hands-on experience, much like becoming an expert in wine or cuisine. The cheesemonger is responsible for all aspects of the cheese inventory: selecting the cheese menu, purchasing, receiving, storage, and ripening.

Production

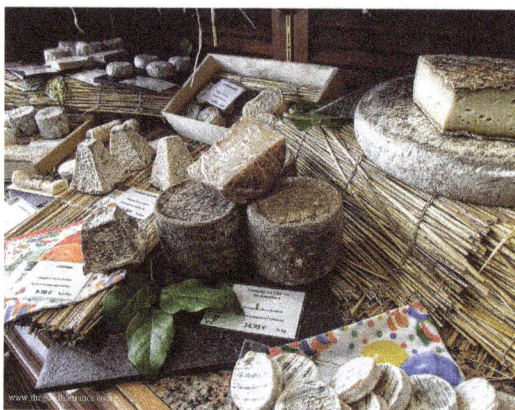

French Cheeses.

In 2014, world production of cheese from whole cow milk was 18.7 million tonnes, with the United States accounting for 29% (5.4 million tonnes) of the world total followed by Germany, France and Italy as major producers.

Other 2014 world totals for processed cheese include:

- From skimmed cow milk, 2.4 million tonnes (leading country, Germany, 845,500 tonnes).

- From goat milk, 523,040 tonnes (leading country, South Sudan, 110,750 tonnes).

- From sheep milk, 680,302 tonnes (leading country, Greece, 125,000 tonnes).

- From buffalo milk, 282,127 tonnes (leading country, Egypt, 254,000 tonnes).

During 2015, Germany, France, Netherlands and Italy exported 10-14% of their produced cheese. The United States was a marginal exporter (5.3% of total cow milk production), as most of its output was for the domestic market.

Consumption

France, Iceland, Finland, Denmark and Germany were the highest consumers of cheese in 2014, averaging 25 kg (55 lb) per person.

Processing

Curdling

During industrial production of Emmental cheese,
the as-yet-undrained curd is broken by rotating mixers.

A required step in cheesemaking is separating the milk into solid curds and liquid whey. Usually this is done by acidifying (souring) the milk and adding rennet. The acidification can be accomplished directly by the addition of an acid, such as vinegar, in a few cases (paneer, queso fresco). More commonly starter bacteria are employed instead which convert milk sugars into lactic acid. The same bacteria (and the enzymes they produce) also play a large role in the eventual flavor of aged cheeses. Most cheeses are made with starter bacteria from the Lactococcus, Lactobacillus, or Streptococcus families. Swiss starter cultures also include Propionibacter shermani, which produces carbon dioxide gas bubbles during aging, giving Swiss cheese or Emmental its holes (called "eyes").

Some fresh cheeses are curdled only by acidity, but most cheeses also use rennet. Rennet sets the cheese into a strong and rubbery gel compared to the fragile curds produced by acidic coagulation alone. It also allows curdling at a lower acidity—important because flavor-making bacteria are inhibited in high-acidity environments. In general, softer, smaller, fresher cheeses are curdled with a greater proportion of acid to rennet than harder, larger, longer-aged varieties.

While rennet was traditionally produced via extraction from the inner mucosa of the fourth stomach chamber of slaughtered young, unweaned calves, most rennet used today in cheesemaking is produced recombinantly. The majority of the applied chymosin is retained in the whey and, at most, may be present in cheese in trace quantities. In ripe cheese, the type and provenance of chymosin used in production cannot be determined.

Curd Processing

At this point, the cheese has set into a very moist gel. Some soft cheeses are now essentially complete: they are drained, salted, and packaged. For most of the rest, the curd is cut into small cubes. This allows water to drain from the individual pieces of curd.

Some hard cheeses are then heated to temperatures in the range of 35–55 °C (95–131 °F). This forces more whey from the cut curd. It also changes the taste of the finished cheese, affecting both the bacterial culture and the milk chemistry. Cheeses that are heated to the higher temperatures are usually made with thermophilic starter bacteria that survive this step—either Lactobacilli or Streptococci.

Salt has roles in cheese besides adding a salty flavor. It preserves cheese from spoiling, draws moisture from the curd, and firms cheese's texture in an interaction with its proteins. Some cheeses are salted from the outside with dry salt or brine washes. Most cheeses have the salt mixed directly into the curds.

Cheese factory in the Netherlands.

Other techniques influence a cheese's texture and flavor. Some examples are:

- Stretching: (Mozzarella, Provolone) The curd is stretched and kneaded in hot water, developing a stringy, fibrous body.

- Cheddaring: (Cheddar, other English cheeses) The cut curd is repeatedly piled up, pushing more moisture away. The curd is also mixed (or *milled*) for a long time, taking the sharp edges off the cut curd pieces and influencing the final product's texture.

- Washing: (Edam, Gouda, Colby) The curd is washed in warm water, lowering its acidity and making for a milder-tasting cheese.

Most cheeses achieve their final shape when the curds are pressed into a mold or form. The harder the cheese, the more pressure is applied. The pressure drives out moisture the molds are designed to allow water to escape and unifies the curds into a single solid body.

Parmigiano-Reggiano in a modern factory.

Ripening

A newborn cheese is usually salty yet bland in flavor and, for harder varieties, rubbery in texture. These qualities are sometimes enjoyed cheese curds are eaten on their own but normally cheeses are left to rest under controlled conditions. This aging period (also called ripening, or, from the French, affinage) lasts from a few days to several years. As a cheese ages, microbes and enzymes transform texture and intensify flavor. This transformation is largely a result of the breakdown of casein proteins and milkfat into a complex mix of amino acids, amines, and fatty acids.

Some cheeses have additional bacteria or molds intentionally introduced before or during aging. In traditional cheesemaking, these microbes might be already present in the aging room; they are simply allowed to settle and grow on the stored cheeses. More often today, prepared cultures are used, giving more consistent results and putting fewer constraints on the environment where the cheese ages. These cheeses include soft ripened cheeses such as Brie and Camembert, blue cheeses such as Roquefort, Stilton, Gorgonzola, and rind-washed cheeses such as Limburger.

Types

There are many types of cheese, with around 500 different varieties recognized by the International Dairy Federation, more than 400 identified by Walter and Hargrove, more than 500 by Burkhalter, and more than 1,000 by Sandine and Elliker. The varieties may be grouped or classified into types according to criteria such as length of ageing, texture, methods of making, fat content, animal milk, country or region of origin, etc. with these criteria either being used singly or in combination, but with no single method being universally used. The method most commonly and traditionally used is based on moisture content, which is then further discriminated by fat content and curing or ripening methods. Some attempts have been made to rationalise the classification of cheese a scheme was proposed by Pieter Walstra which uses the primary and secondary starter combined with moisture content, and Walter and Hargrove suggested classifying by production methods which produces 18 types, which are then further grouped by moisture content.

Moisture content (soft to hard) Categorizing cheeses by firmness is a common but inexact practice. The lines between "soft", "semi-soft", "semi-hard", and "hard" are arbitrary, and many types of cheese are made in softer or firmer variations. The main factor that controls cheese hardness is moisture content, which depends largely on the pressure with which it is packed into molds and aging time.

Fresh, whey and stretched curd cheeses The main factor in the categorization of these cheeses is their age. Fresh cheeses without additional preservatives can spoil in a matter of days.

Content (double cream, goat, ewe and water buffalo) Some cheeses are categorized by the source of the milk used to produce them or by the added fat content of the milk from which they are produced. While most of the world's commercially available cheese is made from cows' milk, many parts of the world also produce cheese from goats and sheep.

Some cheeses are described by their fat content. Double cream cheeses are soft cheeses of cows' milk enriched with cream so that their fat in dry matter (FDM or FiDM) is 60–75%; triple cream cheeses are enriched to at least 75%.

Soft-ripened and blue-vein There are at least three main categories of cheese in which the presence of mold is a significant feature: soft ripened cheese, washed-rind cheese, and blue cheese.

Processed cheeses Processed cheese is made from traditional cheese and emulsifying salts, often with the addition of milk, more salt, preservatives, and food coloring. It is inexpensive, consistent, and melts smoothly. It is sold packaged and either pre-sliced or unsliced, in a number of varieties. It is also available in aerosol cans in some countries.

Bavaria blu cheese.

Edam.

Langres.

Cooking and Eating

At refrigerator temperatures, the fat in a piece of cheese is as hard as unsoftened butter, and its protein structure is stiff as well. Flavor and odor compounds are less easily liberated when cold. For improvements in flavor and texture, it is widely advised that cheeses be allowed to warm up to room temperature before eating. If the cheese is further warmed, to 26–32 °C (79–90 °F), the fats will begin to "sweat out" as they go beyond soft to fully liquid.

Above room temperatures, most hard cheeses melt. Rennet-curdled cheeses have a gel-like protein matrix that is broken down by heat. When enough protein bonds are broken, the cheese itself turns from a solid to a viscous liquid. Soft, high-moisture cheeses will melt at around 55 °C (131 °F), while hard, low-moisture cheeses such as Parmesan remain solid until they reach about 82 °C (180 °F). Acid-set cheeses, including halloumi, paneer, some whey cheeses and many varieties of fresh goat cheese, have a protein structure that remains intact at high temperatures. When cooked, these cheeses just get firmer as water evaporates.

Some cheeses, like raclette, melt smoothly; many tend to become stringy or suffer from a separation of their fats. Many of these can be coaxed into melting smoothly in the presence of acids or starch. Fondue, with wine providing the acidity, is a good example of a smoothly melted cheese dish. Elastic stringiness is a quality that is sometimes enjoyed, in dishes including pizza and Welsh rarebit. Even a melted cheese eventually turns solid again, after enough moisture is cooked off. The saying "you can't melt cheese twice" (meaning "some things can only be done once") refers to the fact that oils leach out during the first melting and are gone, leaving the non-meltable solids behind.

As its temperature continues to rise, cheese will brown and eventually burn. Browned, partially burned cheese has a particular distinct flavor of its own and is frequently used in cooking (e.g., sprinkling atop items before baking them).

Cheeseboard

A cheeseboard (or cheese course) may be served at the end of a meal, either replacing, before or following dessert. The British tradition is to have cheese after dessert, accompanied by sweet wines like Port. In France, cheese is consumed before dessert, with robust red wine. A cheeseboard typically has contrasting cheeses with accompaniments, such as crackers, biscuits, grapes, nuts, celery or chutney. A cheeseboard 70 feet (21 m) long was used to feature the variety of cheeses manufactured in Wisconsin, where the state legislature recognizes a "cheesehead" hat as a state symbol.

Various cheeses on a cheeseboard served with wine for lunch.

Nutrition and Health

The nutritional value of cheese varies widely. Cottage cheese may consist of 4% fat and 11% protein while some whey cheeses are 15% fat and 11% protein, and triple-crème cheeses are 36% fat and 7% protein. In general, cheese is a rich source (20% or more of the Daily Value, DV) of calcium, protein, phosphorus, sodium and saturated fat. A 28-gram (one ounce) serving of cheddar cheese contains about 7 grams (0.25 oz) of protein and 202 milligrams of calcium. Nutritionally, cheese is essentially concentrated milk, but altered by the culturing and aging processes: it takes about 200 grams (7.1 oz) of milk to provide that much protein, and 150 grams (5.3 oz) to equal the calcium.

MacroNutrients (g) of common cheeses per 100 g				
Cheese	Water	Protein	Fat	Carbs
Swiss	37.1	26.9	27.8	5.4
Feta	55.2	14.2	21.3	4.1
Cheddar	36.8	24.9	33.1	1.3
Mozzarella	50	22.2	22.4	2.2
Cottage	80	11.1	4.3	3.4

Vitamin contents in %DV of common cheeses per 100 g													
Cheese	A	B1	B2	B3	B5	B6	B9	B12	Ch.	C	D	E	K
Swiss	17	4	17	0	4	4	1	56	2.8	0	11	2	3
Feta	8	10	50	5	10	21	8	28	2.2	0	0	1	2
Cheddar	20	2	22	0	4	4	5	14	3	0	3	1	3
Mozzarella	14	2	17	1	1	2	2	38	2.8	0	0	1	3
Cottage	3	2	10	0	6	2	3	7	3.3	0	0	0	0

Mineral contents in %DV of common cheeses per 100 g									
Cheese	Ca	Fe	Mg	P	K	Na	Zn	Cu	Mn	Se
Swiss	79	10	1	57	2	8	29	2	0	26
Feta	49	4	5	34	2	46	19	2	1	21
Cheddar	72	4	7	51	3	26	21	2	1	20
Mozzarella	51	2	5	35	2	26	19	1	1	24
Cottage	8	0	2	16	3	15	3	1	0	14

Ch. = Choline; Ca = Calcium; Fe = Iron; Mg = Magnesium; P = Phosphorus; K = Potassium; Na = Sodium; Zn = Zinc; Cu = Copper; Mn = Manganese; Se = Selenium.

All nutrient values including protein are in %DV per 100 g of the food item except for Macro-nutrients.

Cardiovascular Disease

National health organizations, such as the American Heart Association, Association of UK Dietitians, British National Health Service, and Mayo Clinic, among others, recommend that cheese consumption be minimized, replaced in snacks and meals by plant foods, or restricted to low-fat cheeses to reduce caloric intake and blood levels of HDL fat, which is a risk factor for cardiovascular diseases. There is no high-quality clinical evidence that cheese consumption lowers the risk of cardiovascular diseases.

Pasteurization

A number of food safety agencies around the world have warned of the risks of raw-milk cheeses. The U.S. Food and Drug Administration states that soft raw-milk cheeses can cause "serious infectious diseases including listeriosis, brucellosis, salmonellosis and tuberculosis". It is U.S. law since 1944 that all raw-milk cheeses (including imports since 1951) must be aged at least 60 days. Australia has a wide ban on raw-milk cheeses as well, though in recent years exceptions have been made for Swiss Gruyère, Emmental and Sbrinz, and for French Roquefort. There is a trend for cheeses to be pasteurized even when not required by law.

Pregnant women may face an additional risk from cheese; the U.S. Centers for Disease Control has warned pregnant women against eating soft-ripened cheeses and blue-veined cheeses, due to the listeria risk, which can cause miscarriage or harm the fetus.

Fundamentals of Cheese Making

The cheese-making process consists of removing a major part of the water contained in fresh fluid milk while retaining most of the solids. Since storage life increases as water content decreases, cheese making can also be considered a form of food preservation through the process of milk fermentation.

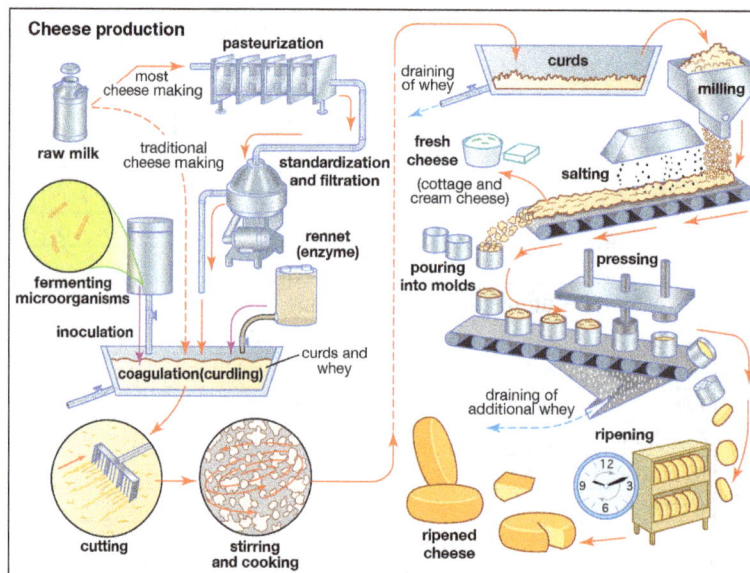

The cheese-making process.

The fermentation of milk into finished cheese requires several essential steps: preparing and inoculating the milk with lactic-acid producing bacteria, curdling the milk, cutting the curd, shrinking the curd (by cooking), draining or dipping the whey, salting, pressing, and ripening. These steps begin with four basic ingredients: milk, microorganisms, rennet, and salt.

Inoculation and Curdling

Milk for cheese making must be of the highest quality. Because the natural microflora present in milk frequently include undesirable types called psychrophiles, good farm sanitation and pasteurization or partial heat treatment are important to the cheese-making process. In addition, the milk must be free of substances that may inhibit the growth of acid-forming bacteria (e.g., antibiotics and sanitizing agents). Milk is often pasteurized to destroy pathogenic microorganisms and to eliminate spoilage and defects induced by bacteria. However, since pasteurization destroys the natural enzymes found in milk, cheese produced from pasteurized milk ripens less rapidly and less extensively than most cheese made from raw or lightly heat-treated milk.

During pasteurization, the milk may be passed through a standardizing separator to adjust the fat-to-protein ratio of the milk. In some cases the cheese yield is improved by concentrating protein in a process known as ultrafiltration. The milk is then inoculated with fermenting microorganisms and rennet, which promote curdling.

The fermenting microorganisms carry out the anaerobic conversion of lactose to lactic acid. The type of organisms used depends on the variety of cheese and on the production process. Rennet is an enzymatic preparation that is usually obtained from the fourth stomach of calves. It contains a number of proteolytic (protein-degrading) enzymes, including rennin and pepsin. Some cheeses, such as cottage cheese and cream cheese, are produced by acid coagulation alone. In the presence of lactic acid, rennet, or both, the milk protein casein clumps together and precipitates out of solution; this is the process known as curdling, or coagulation. Coagulated casein assumes a solid or gellike structure (the curd), which traps most of the fat, bacteria, calcium, phosphate, and other particulates. The remaining liquid (the whey) contains water, proteins resistant to acidic and enzymatic denaturation (e.g., antibodies), carbohydrates (lactose), and minerals.

Lactic acid produced by the starter culture organisms has several functions. It promotes curd formation by rennet (the activity of rennet requires an acidic pH), causes the curd to shrink, enhances whey drainage (syneresis), and helps prevent the growth of undesirable microorganisms during cheese making and ripening. In addition, acid affects the elasticity of the finished curd and promotes fusion of the curd into a solid mass. Enzymes released by the bacterial cells also influence flavour development during ripening.

Salt is usually added to the curd. In addition to enhancing flavour, it helps to withdraw the whey from the curd and inhibits the growth of undesirable microorganisms.

Cutting and Shrinking

After the curd is formed, it is cut with fine wire "knives" into small cubes approximately one centimetre (one-half inch) square. The curd is then gently heated, causing it to shrink. The degree of shrinkage determines the moisture content and the final consistency of the cheese. Whey is removed by draining or dipping. The whey may be further processed to make whey cheeses (e.g., ricotta) or beverages, or it may be dried in order to preserve it as a food ingredient.

Cheese curds, Cheese curds rolling along a conveyor belt
in preparation for being cut, stirred, and cooked.

Ripening

Most cheese is ripened for varying amounts of time in order to bring about the chemical changes necessary for transforming fresh curd into a distinctive aged cheese. These changes are catalyzed by enzymes from three main sources: rennet or other enzyme preparations of animal or vegetable origin added during coagulation, microorganisms that grow within the cheese or on its surface, and the cheese milk itself. The ripening time may be as short as one month, as for Brie, or a year or more, as in the case of sharp cheddar.

The ripening of cheese is influenced by the interaction of bacteria, enzymes, and physical conditions in the curing room. The speed of the reactions is determined by temperature and humidity conditions in the room as well as by the moisture content of the cheese. In most cheeses lactose continues to be fermented to lactic acid and lactates, or it is hydrolyzed to form other sugars. As a result, aged cheeses such as Emmentaler and cheddar have no residual lactose.

In a similar manner, proteins and lipids (fats) are broken down during ripening. The degree of protein decomposition, or proteolysis, affects both the flavour and the consistency of the final cheese. It is especially apparent in Limburger and some blue-mold ripened cheeses. Surface-mold ripened cheeses, such as Brie, rely on enzymes produced by the white Penicillium camemberti mold to break down proteins from the outside. When lipids are broken down (as in Parmesan and Romano cheeses), the process is called lipolysis.

The eyes, or holes, typical of Swiss-type cheeses such as Emmentaler and Gruyère come from a secondary fermentation that takes place when, after two weeks, the cheeses are moved from refrigerated curing to a warmer room, where temperatures are in the range of 20 to 24 °C (68 to 75 °F). At this stage, residual lactates provide a suitable medium for propionic acid bacteria (Propionibacterium shermanii) to grow and generate carbon dioxide gas. Eye formation takes three to six weeks. Warm-room curing is stopped when the wheels develop a rounded surface and the echo of holes can be heard when the cheese is thumped. The cheese is then moved back to a cold room, where it is aged at about 7 °C (45 °F) for 4 to 12 months in order to develop its typical sweet, nutty flavour.

The unique ripening of blue-veined cheeses comes from the mold spores Penicillium roqueforti or P. glaucum, which are added to the milk or to the curds before pressing and are activated by air. Air is introduced by "needling" the cheese with a device that punches about 50 small holes into the top. These air passages allow mold spores to grow vegetative cells and spread their greenish blue myce-lia, or threadlike structures, through the cheese. Penicillium molds are also rich in proteolytic and lipolytic enzymes, so that during ripening a variety of trace compounds also are produced, such as free amines, amino acids, carbonyls, and fatty acids all of which ultimately affect the flavour and texture of the cheese.

Surface-ripened cheeses such as Gruyère, brick, Port Salut, and Limburger derive their flavour from both internal ripening and the surface environment. For instance, the high-moisture wip-ing of the surface of Gruyère gives that cheese a fuller flavour than its Emmentaler counterpart. Specific organisms, such as Brevibacterium linens, in Limburger cheese result in a reddish brown surface growth and the breakdown of protein to amino nitrogen. The resulting odour is offensive to some, but the flavour and texture of the cheese are pleasing to many.

Not all cheeses are ripened. Cottage, cream, ricotta, and most mozzarella cheeses are ready for sale as soon as they are made. All these cheeses have sweet, delicate flavours and often are combined with other foods.

Varieties of Cheese

As a result of the many combinations of milks, cultures, enzymes, molds, and technical processes, literally hundreds of varieties of cheese are made throughout the world. The different types of cheese can be classified in many ways; the most effective is probably according to hardness or rip-ening method. The table groups several varieties of cheese based on these criteria.

Varieties of cheese, classified by hardness and ripening method		
	Ripening method	cheese variety
Very hard	Bacteria/enzymes	Asiago, Parmesan, Romano, Sapsago, Sonoma Dry Jack
Hard	Bacteria/enzymes	Cantal, cheddar, Colby
	Eye-producing bacteria/enzymes	Emmentaler (Swiss), Gruyère, Fontina, Jarlsberg
Semihard/semi-soft	bacteria/enzymes	Brick, Edam, Gouda, Monterey Jack, mozzarella, Munster, provolone
	Bacteria/enzymes and surface mi-croorganisms	Bel Paese, brick, Limburger, Port Salut, Trappist
	Bacteria/enzymes and blue mold	Blue, Gorgonzola, Roquefort, Stilton
Soft	Bacteria/enzymes and surface mi-croorganisms	Brie, Camembert, Neufchâtel (France), Pont l'Évêque
	Unripened	Baker's, cottage, cream, feta, Neufchâtel (United States), pot

In recent years different types of cheese have been combined in order to increase variety and consumer interest. For example, soft and mildly flavoured Brie is combined with a more pungent semisoft cheese such as blue or Gorgonzola. The resulting "Blue-Brie" has a bloomy white edible rind, while its interior is marbled with blue Penicillium roqueforti mold. The cheese is marketed under various names such as Bavarian Blue, Cambazola, Lymeswold, and Saga Blue. Another combination cheese is Norwegian Jarlsberg. This cheese results from a marriage of the cultures and manufacturing procedures for Dutch Gouda and Swiss Emmentaler.

Pasteurized Process Cheese

Some natural cheese is made into process cheese, a product in which complete ripening is halted by heat. The resulting product has an indefinite shelf life. Most process cheese is used in food service outlets and other applications where convenient, uniform melting is required.

Pasteurized process cheese is made by grinding and mixing natural cheese with other ingredients, such as water, emulsifying agents, colouring, fruits, vegetables, or meat. The mixture is then heated to temperatures of 74 °C (165 °F) and stirred into a homogeneous, plastic mass. Process cheese foods, spreads, and products differ from process cheese in that they may contain other ingredients, such as nonfat dry milk, cheese whey, and whey protein concentrates, as well as additional amounts of water.

American cheddar is processed most frequently. However, other cheeses such as washed-curd, Colby, Swiss, Gruyère, and Limburger are similarly processed. In a slight variation, cold pack or club cheese is made by grinding and mixing together one or more varieties of cheese without heat. This cheese food may contain added flavours or ingredients.

FERMENTED MILK

The primary function of fermenting milk was, originally, to extend its shelf life. With this came numerous advantages, such as an improved taste and enhanced digestibility of the milk, as well as the manufacture of a wide variety of products. Historically the fermentation of milk can be traced back to around 10 000 B.C. It is likely that fermentation initially arose spontaneously from indigenous microflora found in milk. Fortunately, the bacteria were lactococci and lactobacilli which typically suppress spoilage and pathogenic organisms effectively. The evolution of these products likely came as a result of the climate of the region in which they were produced: thermophilic lactic acid fermentation favours the heat of the sub-tropics; mesophilic lactic acid fermentation occurs at cooler temperatures. Today the fermentations are controlled with specific starter cultures and conditions. Some of the many fermented milk products are: acidophilus milk, crème fraîche, cultured buttermilk, kefir, koumiss, filmjölk, sour cream, and viili. Yogurt and cheese are also fermented milk products.

Fermented milk products can be classified into 3 categories:

- Viscous products,

- Beverage products,

- Carbonated products.

Within these categories, the fermented milk products may be fresh, or have an extended shelf life. The fresh products contain live starter culture bacteria, including probiotics, while the extended shelf life products contain no live microorganisms.

Product	Typical Shelf Life(4 °C)
Acidophilus Milk	2 wks
Cultured Buttermilk	10 d
Sour Cream	4 wks
Kefir	10-14 d
Koumiss	10-14 d
Filmjölk	10-14 d
Viili	14 d
Crème Fraîche	10 d

There are numerous factors which affect the outcome of the product including the chemical composition of the milk, additives and starter cultures used, as well as the processing of the product. They affect the ultimate flavour, texture, and consistency of the final product. It is not uncommon for the manufacturer to add stabilizers such as pectins and gums, in order to avoid the sedimentation of milks solids and the separation of whey in the package, while improving the mouthfeel of the product.

The general process by which fermented milk products are made begins with a preliminary treatment of milk which may include clarification, fat separation and standardization, and evaporation. Processing follows next, with de-aeration, homogenization, and pasteurization. The milk is then cooled to the appropriate fermentation temperature and starter cultures are added.

Starter cultures differ for each product. They consist of microorganisms added to the milk to provide specific characteristics in the finished fermented milk product in a controlled and predictable manner. The primary function of lactic acid starters is to ferment lactose into lactic acid, but they may also contribute to flavour, aroma and alcohol production, while inhibiting spoilage microorganisms. A single strain of bacteria may be added, or a mixture of several microorganisms may be introduced. The bacteria, yeasts and moulds work at different temperatures as well. Thermophilic lactic acid fermentation favour hot temperatures (40-45 °C) while mesophilic lactic acid fermentation occurs at cooler temperatures (25 and 40 °C).

As the starter cultures grow within the milk, fermentation takes place. Fermentation is the chemical conversion of carbohydrates into alcohols or acids. In fermented milk products both alcohol and lactic acid may be produced, like in kefir and koumiss, or just lactic acid, like in sour cream. The bacteria ingest the lactose (milk sugar), and release lactic acid as waste causing the acidity to increase. This rise in acidity causes the milk proteins to denature (unfold) and tangle themselves into masses (curds) while also inhibiting the growth of other organisms that are not acid tolerant. Following the completion of fermentation, flavourings can be added and the products are packaged, labeled and put into cold storage before being sent to stores.

Varieties

- Kefir is made most often from partially skimmed cow's milk. It can be packaged either as natural or plain kefir with no added fruit or flavours or as flavoured kefir. The final product contains live bacteria and yeasts that produce carbon dioxide gas. This gas production gives kefir a "sparkling" sensation on the tongue when eaten. Kefir has been referred to as the champagne of fermented dairy products.

- Koumiss: Mare's milk has higher sugar content than cow's and goat's milk, and as a result koumiss has a slightly higher alcohol content than kefir. Today, cow's milk is generally used for koumiss, with the addition of sugar to better approximate the composition of mare's milk.

- Cultured Buttermilk may contain added butterflakes, fruit condiments, or flavourings. It is also available with different fat contents.

- Viili comes in a wide range of varieties, including products of different fat content,lactose-reduced varieties and flavoured versions. Viili can be made from homogenizedmilk and without mould growing on the surface.

- Sour cream comes in full fat (minimum 14% fat), low fat and fat free varieties.

- Filmjölk has fruit flavoured variants and can have the addition of beneficial probiotic bacteria such as Bifidobacterium lactis and many species of lactobacilli.

Composition

Based on 100 g	Kefir (100 g)			Sour Cream		
	2%	0%	3.80%	Full Fat	Low Fat	Fat-Free
Fat	2.0 g	0 g	3.2 g	20.1 g	10.6 g	0 g
Protein	3.4 g	3.2 g	3.2 g	3.1 g	3.5 g	3.1 g
Carbohydrates	4.6 g	4.8 g	7.2 g	4.3 g	7.1 g	15.6 g
Sugar	3.4 g	4 g	4 g	0.16 g	0.22 g	0.39 g
Calcium	220 mg	165 mg	165 mg	116 mg	141 mg	125 mg

Based on 100 g	Cultrured Buttermilk		Acidophilus milk (2%)	Crème Fraiche
	Low Fat	Fat Free		
Fat	2.0 g	0.88 g	3.8 g	38.7 g
Protein	4.1 g	3.3 g	1 g	1.8 g
Carbohydrates	5.3 g	4.8 g	5.4 g	2.5 g
Sugar	5.3 g	4.8 g	5 g	2.5 g
Calcium	143 mg	116 mg	125 mg	60 mg

Based on 100 g	Koumiss (mare's milk)	Koumiss (cow's milk)
Fat	1.1 g	3.9 g
Protein	3.5 g	3.3 g
Carbohydrates	6.1 g	4.7 g
Sugar	6.1 g	4.7 g
Calcium	90 mg	~120 mg

Viili and filmjölk likely have compositions similar to kefir depending on the level of milk fat used to prepare them.

Various Uses

Kefir and Koumiss can be used in smoothies, salad dressings, and sauces. They can be added to baked goods such as pancakes, waffles, and breads, or in soups and desserts as a replacement for other milk products such as yogurt or buttermilk. They are also delicious mixed with fresh fruit or cereal as a breakfast, lunch or snack. They all make refreshing beverages on their own or mixed with fruits, honey, maple syrup, iced coffees and teas as well as other sweeteners and flavours.

Filmjölk is eaten in the same way as yogurt, usually from a bowl using a spoon. It is sometimes drunk as a thick beverage. Many people add sugar, jam, applesauce, cinnamon or berries. Cereals, corn flakes or muesli are often added to filmjölk. In northern regions of Sweden, crushed crisp bread is sometimes put into it. It could be used in smoothies, salad dressings and sauces, as well as in baked goods.

Cultured Buttermilk is a versatile ingredient in baking. It works very well in biscuits, breads, and desserts. Cultured buttermilk is often used in salad dressings and sauces, stirred into mashed potatoes and soups and it has even been used to make tangy buttermilk ice cream. It is also considered a refreshing beverage.

Sour Cream has numerous applications. It is commonly used as a base for dips, salad dressings and sauces. It is eaten as a condiment on potatoes, chili, or with smoked salmon, as well as many other foods. Sour cream can be used in soups and works well in baked products like breads, cakes, pies and cookies. Sour cream has significantly less calories than mayonnaise and performs many similar functions. In Russian cuisine, sour cream is often added to borscht and other soups. In Tex-Mex cuisine, it is often added to tacos, nachos, burritos, taquitos or guacamole. Hungarian cooks use it as an ingredient in sauces and in recipes such as ham-filled crepes.

Crème Fraîche has similar uses to sour cream, however it's sweeter flavour makes it particularly well suited to desserts, as a topping, or as a base for other flavours. Crème fraîche works well in dips, dressings and sauces or as an addition to soups.

Viili is consumed fresh and chilled, in the same way as yogurt, and can be topped with fruits, nuts, or cereal, as well as other flavourings like spices, sugar and honey. Viili may be added to smoothies, or used in baked goods. It may also be flavoured like yogurt.

Acidophilus milk is consumed as a beverage by the glass, or added to cereal. It has been used to make egg nog. Acidophilus milk also works well in sauces and desserts.

Functional Properties

Fermented milk products have numerous functional properties:

- Preservation: Bacteria are inhibited from growing through pH reduction when lactic acid is formed, and shelf life is increased.

- Flavour Enhancement: The sour characteristic of fermented milk products comes from fermentation products (lactic acid, diactyl, carbon dioxide, ethanol); these products act as excellent flavour carriers for herbs, spices and other flavourings.

- Texture Enhancement: Some fermented milk products (sour cream or crème fraîche) can add body and thickness to sauces, dips or vinaigrettes.

- Reducing Caloric Content: Many fermented milk products come in low fat or fat free varieties and can be used to substitute for higher fat ingredients.

- Emulsification: Milk proteins help stabilize fat emulsions in salad dressings, soups and cakes.

- Foaming and Whipping: crème fraîche is capable of being whipped like whip cream.

- Nutritional benefits: Fermented milk products may contain probiotics (bacteria that are beneficial to health) as well as many vitamins and minerals.

CLABBER

Clabber milk is a naturally fermented milk product that can be eaten raw or used in recipes. It also has a little leavening power all on its own so it's great to add to baked goods. Raw cow's milk is full of naturally occurring beneficial Lactic Acid Bacteria and when that bacteria is supported with a warm environment it will ferment the milk creating something similar to a cross between yogurt and kefir. Eventually, if left to ferment long enough the clabber milk will separate into curds and whey.

Fermenting or souring milk is very different than having milk spoil. Spoiled milk only occurs if the beneficial bacteria found in clean raw cow's milk has been killed by pasteurization thus allowing mold spores or other contaminants to flourish. In a fermented milk product the Lactic Acid Bacteria have soured the milk with the lactic acids they produce while consuming lactose. The higher acidity of the souring process keeps other microbes (that can be harmful to humans) from forming. It is very important that you use only high quality raw milk from clean grass fed cows when making clabber milk.

Clabber milk with honey.

Making Clabber Milk

Ferment your raw milk in a clean glass jar with a clean loose fitting lid at room temperature until the milk sours and starts to separate. This can take between 1-5 days depending on the age of the milk, the temperature in your home and the natural bacteria in the milk itself.

When the clabber has solidified it can then be skimmed of the clotted cream, used for baking, eaten like yogurt, or it can be strained to separate the curds from the whey.

A bubbly jar of Clabber.

A cotton bag for separating the curds and whey.

The golden whey draining from the curds.

After straining the clabber the whey can be used as a starter for any lacto-fermented project from veggies to grains and is especially useful for starting a new batch of clabber milk. Using a tablespoon of clabber whey in the new batch of milk will speed the fermentation process along considerably. The curds will thicken and sweeten with straining and take on a cream cheese like texture.

SOURED MILK

Soured milk denotes a range of food products produced by the acidification of milk. Acidification, which gives the milk a tart taste, is achieved either through bacterial fermentation or through the addition of an acid, such as lemon juice or vinegar. The acid causes milk to coagulate and thicken, inhibiting the growth of harmful bacteria and improving the product's shelf life.

Soured milk that is produced by bacterial fermentation is more specifically called fermented milk or cultured milk. Traditionally, soured milk was simply fresh milk that was left to ferment and sour by keeping it in a warm place for a day, often near a stove. Modern commercial soured milk may differ from milk that has become sour naturally.

Soured milk that is produced by the addition of an acid, with or without the addition of microbial organisms, is more specifically called acidified milk. In the United States, acids used to manufacture acidified milk include acetic acid (commonly found in vinegar), adipic acid, citric acid (commonly found in lemon juice), fumaric acid, glucono-delta-lactone, hydrochloric acid, lactic acid, malic acid, phosphoric acid, succinic acid, and tartaric acid.

Soured milk is commonly made at home or is sold and consumed in Europe, especially in Eastern Europe (Bulgaria, Belarus, Poland, Slovakia, Russia, Ukraine), all over the countries of the former Yugoslavia (Macedonia, Serbia, Montenegro, Bosnia and Herzegovina, Croatia, Slovenia), Romania, Greece, Finland, Germany, and Scandinavia.

It is also made at home or sold in supermarkets and consumed in the Great Lakes region of Somalia and Eastern Africa (Kenya, Uganda, Rwanda, Burundi and Tanzania). It is also a traditional food of the Bantu people of Southern Africa.

Since the 1970s, some producers have used chemical acidification in place of biological agents.

In Recipes

Raw milk that has not gone sour is sometimes referred to as "sweet milk", because it contains the sugar lactose. Fermentation converts the lactose to lactic acid, which has a sour flavor. Before the invention of refrigeration, raw milk commonly became sour before it could be consumed, and various recipes incorporate such leftover milk as an ingredient. Sour milk produced by fermentation differs in flavor from that produced by acidification, because the acids commonly added in commercial manufacture have different flavors from lactic acid, and also because fermentation can introduce new flavors. Buttermilk is a common modern substitute for naturally soured milk.

KEFIR

Kefir or kephir, is a fermented milk drink similar to a thin yogurt that is made from kefir grains, a specific type of mesophilic symbiotic culture. The drink originated in the North Caucasus, Eastern Europe and Russia, where it is prepared by inoculating cow, goat, or sheep milk with kefir grains.

Kefir grains, a symbiotic matrix of bacteria and yeasts.

Traditional kefir is fermented at ambient temperatures, generally overnight. Fermentation of the lactose yields a sour, carbonated, slightly alcoholic beverage, with a consistency and taste similar to drinkable yogurt.

The kefir grains initiating the fermentation consist of a symbiotic culture of lactic acid bacteria and yeasts embedded in a matrix of proteins, lipids, and polysaccharides. The matrix is formed by microbial activity and resemble small cauliflower grains, with color ranging from white to creamy yellow. A complex and highly variable community can be found in these grains, which can include lactic acid bacteria, acetic acid bacteria, and yeasts. While some microbes predominate, Lactobacillus species are always present. The microbe flora can vary between batches of kefir due to factors such as the kefir grains rising out of the milk while fermenting or curds forming around the grains, as well as temperature.

During fermentation, changes in the composition of ingredients occur. Lactose, the sugar present in milk, is broken down mostly to lactic acid (25%) by the lactic acid bacteria, which results in acidification of the product. Propionibacteria further break down some of the lactic acid into propionic acid (these bacteria also carry out the same fermentation in Swiss cheese). Other substances that contribute to the flavor of kefir are pyruvic acid, acetic acid, diacetyl and acetoin (both of which contribute a "buttery" flavor), citric acid, acetaldehyde, and amino acids resulting from protein breakdown.

Low Lactose Content

The slow-acting yeasts, late in the fermentation process, break lactose down into ethanol and carbon dioxide. As a result of the fermentation, very little lactose remains in kefir. People with lactose intolerance are able to tolerate kefir, provided the number of live bacteria present in this beverage consumed is high enough (i.e., fermentation has proceeded for adequate time). It has also been shown that fermented milk products have a slower transit time than milk, which may further improve lactose digestion.

Alcohol/Ethanol Content

Kefir contains ethanol, which is detectable in the blood of human consumers. The level of ethanol in kefir can vary by production method. A study of kefir sold in Germany showed an ethanol level of only 0.02 g per litre, which was attributed to fermentation under controlled conditions allowing the growth of *Lactobacteria* only, but excluding the growth of other microorganisms that form much higher amounts of ethanol. A study of German commercial kefir found levels of 0.002-0.005% of ethanol. Another study found levels of ethanol of 2.10%, 1.46% and 1.40% in cow, goat and sheep kefir, respectively. Kefir produced by small-scale dairies in Russia early in the 20th century had 1-2% ethanol. Modern processes, which use shorter fermentation times, result in much lower ethanol concentrations of 0.2−0.3%.

Nutrition

Composition

Kefir products contain nutrients in varying amounts from negligible to significant, including dietary minerals, vitamins, essential amino acids, and conjugated linoleic acid, in amounts similar to unfermented cow, goat, or sheep milk. At a pH of 4.2 - 4.6, kefir is composed mainly of water and by-products of the fermentation process, including carbon dioxide and ethanol.

Typical of milk, several dietary minerals are found in kefir, such as calcium, iron, phosphorus, magnesium, potassium, sodium, copper, molybdenum, manganese, and zinc in amounts that have not been standardized to a reputable nutrient database. Also similar to milk, kefir contains vitamins in variable amounts, including vitamin A, vitamin B_1 (thiamine), vitamin B_2 (riboflavin), vitamin B_3 (niacin), vitamin B_6 (pyridoxine), vitamin B_9 (folic acid), vitamin B_{12} (cyanocobalamin), vitamin C, vitamin D, and vitamin E. Essential amino acids found in kefir include methionine, cysteine, tryptophan, phenylalanine, tyrosine, leucine, isoleucine, threonine, lysine, and valine, as for any milk product.

Microbiota

Probiotic bacteria found in kefir products include: Lactobacillus acidophilus, Bifidobacterium bifidum, Streptococcus thermophilus, Lactobacillus delbrueckii subsp. bulgaricus, Lactobacillus helveticus, Lactobacillus kefiranofaciens, Lactococcus lactis, and Leuconostoc species. Lactobacilli in kefir may exist in concentrations varying from approximately 1 million to 1 billion colony-forming units per milliliter, and are the bacteria responsible for the synthesis of the polysaccharide kefiran.

In addition to bacteria, kefir often contains strains of yeast that can metabolize lactose, such as Kluyveromyces marxianus, Kluyveromyces lactis, and Saccharomyces fragilis, as well as strains of yeast that do not metabolize lactose, including Saccharomyces cerevisiae, Torulaspora delbrueckii, and Kazachstania unispora. The nutritional significance of these strains is unknown.

Production

90 grams of kefir grains.

Kefir production.

Kefir is made by adding kefir grains to milk typically at a proportion of 2-5% grains-to-milk. The mixture is then placed in a corrosion-resistant container, such as a glass jar, and stored preferably in the dark to prevent degradation of light-sensitive vitamins. After a period between 12-24 hours of fermentation at mild temperature, ideally 20–25 °C (68–77 °F), the grains are strained from the milk using a corrosion-resistant (stainless steel or plastic) utensil and kept to produce another batch. During the fermentation process the grains enlarge and eventually split forming new units.

The resulting fermented liquid, may be drunk, used in recipes, or kept aside in a sealed container for additional time to undergo a secondary fermentation. Because of its acidity the beverage should not be stored in reactive metal containers such as aluminium, copper, or zinc, as these may leach into it over time. The shelf life, unrefrigerated, is up to thirty days.

The Russian method permits production of kefir on a larger scale and uses two fermentations. The first step is to prepare the cultures by inoculating milk with 2–3% grains as described. The grains are then removed by filtration and 1–3% of the resulting liquid mother culture is added to milk and fermented for 12 to 18 hours.

Kefir can be made using freeze-dried cultures commonly available in powder form from health food stores. A portion of the resulting kefir can be saved to be used a number of times to propagate further fermentations but ultimately does not form grains.

In Taiwan, researchers were able to produce kefir in laboratory using microorganisms isolated from kefir grains. They report that the resulting kefir drink had chemical properties similar to homemade kefir.

Milk Types

Kefir grains will ferment the milk from most mammals and will continue to grow in such milk. Typical animal milks used include cow, goat, and sheep, each with varying organoleptic (flavor, aroma, and texture) and nutritional qualities. Raw milk has been traditionally used.

Kefir grains will also ferment milk substitutes such as soy milk, rice milk, nut milk and coconut milk, as well as other sugary liquids including fruit juice, coconut water, beer wort, and ginger beer. However, the kefir grains may cease growing if the medium used does not contain all the growth factors required by the bacteria.

Milk sugar is not essential for the synthesis of the polysaccharide that makes up the grains (kefiran), and rice hydrolysate is a suitable alternative medium. Additionally, kefir grains will reproduce when fermenting soy milk, although they will change in appearance and size due to the differing proteins available to them.

A variation of kefir grains that thrive in sugary water also exists, and can vary markedly from milk kefir in both appearance and microbial composition.

Culinary

Lithuanian kefir-based borscht.

As it contains Lactobacillus bacteria, kefir can be used to make a sourdough bread. It is also useful as a buttermilk substitute in baking. Kefir is one of the main ingredients in borscht in Lithuania, also known in Poland as Lithuanian cold soup (chłodnik litewski), and other countries. Kefir-based soup okroshka is common across the former Soviet Union. Kefir may be used in place of milk on cereal, granola, milkshakes, salad dressing, ice cream, smoothies and soup.

Kefir is a fermented drink, traditionally made using cow's milk or goat's milk.

It is made by adding kefir grains to milk. These are not cereal grains, but grain-like colonies of yeast and lactic acid bacteria that resemble a cauliflower in appearance.

Over approximately 24 hours, the microorganisms in the kefir grains multiply and ferment the sugars in the milk, turning it into kefir.

Then the grains are removed from the liquid and can be used again.

In other words, kefir is the drink, but kefir grains are the starter culture that you use to produce the beverage.

Kefir originated from parts of Eastern Europe and Southwest Asia. The name is derived from the Turkish word keyif, which means "feeling good" after eating.

The grains lactic acid bacteria turn the milk's lactose into lactic acid, so kefir tastes sour like yogurt but has a thinner consistency.

A 6-ounce (175-ml) serving of low-fat kefir contains:

- Protein: 4 grams.

- Calcium: 10% of the RDI.

- Phosphorus: 15% of the RDI.

- Vitamin B12: 12% of the RDI.

- Riboflavin (B2): 10% of the RDI.

- Magnesium: 3% of the RDI.

- A decent amount of vitamin D.

In addition, kefir has about 100 calories, 7–8 grams of carbs and 3–6 grams of fat, depending on the type of milk used.

Kefir also contains a wide variety of bioactive compounds, including organic acids and peptides that contribute to its health benefits.

Dairy-free versions of kefir can be made with coconut water, coconut milk or other sweet liquids. These will not have the same nutrient profile as dairy-based kefir.

Kefir is a more Powerful Probiotic than Yogurt

Some microorganisms can have beneficial effects on health when ingested. Known as probiotics, these microorganisms may influence health in numerous ways, aiding digestion, weight management and mental health.

- Yogurt is the best known probiotic food in the Western diet, but kefir is actually a much more potent source.

- Kefir grains contain up to 61 strains of bacteria and yeasts, making them a very rich and diverse probiotic source, though diversity may vary.

Other fermented dairy products are made from far fewer strains and don't contain any yeasts.

Kefir has Potent Antibacterial Properties

Certain probiotics in kefir are believed to protect against infections:

- This includes the probiotic Lactobacillus kefiri, which is unique to kefir.

- Studies demonstrate that this probiotic can inhibit the growth of various harmful bacteria, including Salmonella, Helicobacter pylori and E. coli.

- Kefiran, a type of carbohydrate present in kefir, also has antibacterial properties.

- Kefir Can Improve Bone Health and Lower the Risk of Osteoporosis.

- Osteoporosis is characterized by deterioration of bone tissue and is a major problem in Western countries.

- It is especially common among older women and dramatically raises your risk of fractures.

- Ensuring an adequate calcium intake is one of the most effective ways to improve bone health and slow the progression of osteoporosis.

- Full-fat kefir is not only a great source of calcium but also vitamin K2 — which plays a central role in calcium metabolism. Supplementing with K2 has been shown to reduce your risk of fractures by as much as 81%.

- Recent animal studies link kefir to increased calcium absorption in bone cells. This leads to improved bone density, which should help prevent fractures.

Kefir May be Protective against Cancer

Cancer is one of the world's leading causes of death:

- It occurs when abnormal cells in your body grow uncontrollably, such as in a tumor.

- The probiotics in fermented dairy products are believed to reduce tumor growth by stimulating your immune system. Therefore, it is possible that kefir may fight cancer.

- This protective role has been demonstrated in several test-tube studies.

- One study found that kefir extract reduced the number of human breast cancer cells by 56%, compared to only 14% for yogurt extract.

- Keep in mind that human studies are needed before firm conclusions can be made.

The Probiotics in it may Help with Various Digestive Problems

Probiotics such as kefir can help restore the balance of friendly bacteria in your gut:

- This is why they are highly effective at treating many forms of diarrhea.

- What's more, ample evidence suggests that probiotics and probiotic foods can alleviate many digestive problems.

- These include irritable bowel syndrome (IBS), ulcers caused by H. pylori infection and many others.

- For this reason, kefir may be useful if you have problems with digestion.

Kefir is Low in Lactose

Regular dairy foods contain a natural sugar called lactose:

- Many people, especially adults, are unable to break down and digest lactose properly. This condition is called lactose intolerance.

- The lactic acid bacteria in fermented dairy foods like kefir and yogurt turn the lactose into lactic acid, so these foods are much lower in lactose than milk.

- They also contain enzymes that can help break down the lactose even further.

- Therefore, kefir is generally well tolerated by people with lactose intolerance, at least compared to regular milk.

- Keep in mind that it is possible to make kefir that is 100% lactose-free by using coconut water, fruit juice or another non-dairy beverage.

Kefir may Improve Allergy and Asthma Symptoms

Allergic reactions are caused by inflammatory responses against certain foods or substances:

- People with an over-sensitive immune system are more prone to allergies, which can provoke conditions like asthma.

- In animal studies, kefir has been shown to suppress inflammatory responses related to allergies and asthma.

- Human studies are needed to better explore these effects.

KUMIS

Kumis (also spelled kumiss or koumiss or kumys,) is a fermented dairy product traditionally made from mare's milk. The drink remains important to the peoples of the Central Asian steppes, of Huno-Bulgar, Turkic and Mongol origin: Kazakhs, Bashkirs, Kalmyks, Kyrgyz, Mongols, and Yakuts.

Kumis is a dairy product similar to kefir, but is produced from a liquid starter culture, in contrast to the solid kefir "grains". Because mare's milk contains more sugars than cow's or goat's milk, when fermented, kumis has a higher, though still mild, alcohol content compared to kefir.

Even in the areas of the world where kumis is popular today, mare's milk remains a very limited commodity. Industrial-scale production, therefore, generally uses cow's milk, which is richer in fat and protein, but lower in lactose than the milk from a horse. Before fermentation, the cow's milk is fortified in one of several ways. Sucrose may be added to allow a comparable fermentation. Another technique adds modified whey to better approximate the composition of mare's milk.

Production of Mare's Milk

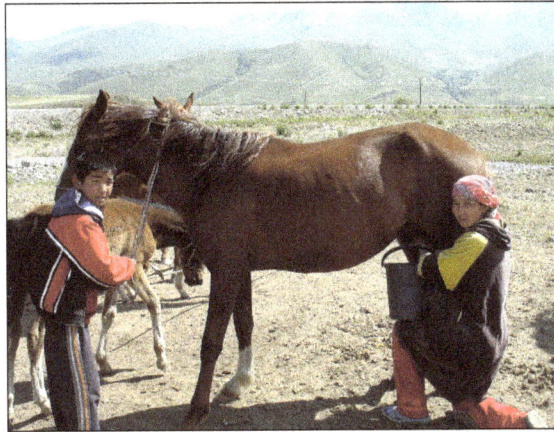

A mare being milked in the Suusamyr Valley, Kyrgyzstan.

A 1982 source reported 230,000 horses were kept in the Soviet Union specifically for producing milk to make into kumis. Rinchingiin Indra, writing about Mongolian dairying, says "it takes considerable skill to milk a mare" and describes the technique: the milker kneels on one knee, with a pail propped on the other, steadied by a string tied to an arm. One arm is wrapped behind the mare's rear leg and the other in front. A foal starts the milk flow and is pulled away by another person, but left touching the mare's side during the entire process.

In Mongolia, the milking season for horses traditionally runs between mid-June and early October. During one season, a mare produces approximately 1,000 to 1,200 litres of milk, of which about half is left to the foals.

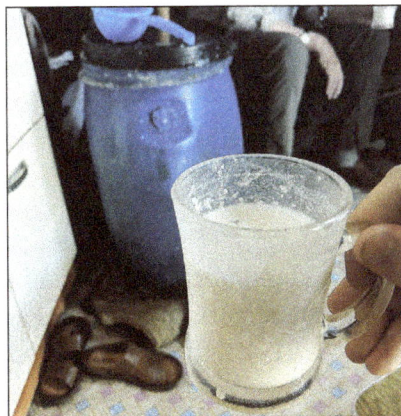

A glass of homemade Mongolian airag, prepared
in the blue plastic barrel in the background.

Kumis is made by fermenting raw unpasteurized mare's milk over the course of hours or days, often while stirring or churning. The physical agitation has similarities to making butter. During the fermentation, lactobacilli bacteria acidify the milk, and yeasts turn it into a carbonated and mildly alcoholic drink.

Traditionally, this fermentation took place in horse-hide containers, which might be left on the top of a yurt and turned over on occasion, or strapped to a saddle and joggled around over the course of a day's riding. Today, a wooden vat or plastic barrel may be used in place of the leather container.

In modern controlled production, the initial fermentation takes two to five hours at a temperature of around 27 °C (81 °F); this may be followed by a cooler aging period. The finished product contains between 0.7 and 2.5% alcohol.

Kumis itself has a very low level of alcohol, comparable to small beer, the common drink of medieval Europe that also helps to avoid the consumption of potentially contaminated water. Kumis can, however, be strengthened through freeze distillation, a technique Central Asian nomads are reported to have employed. It can also be made into the distilled beverage known as araka or arkhi.

Consumption

Kumis-flavored ice cream at a restaurant in Nur-Sultan, Kazakhstan.

Strictly speaking, *kumis* is in its own category of alcoholic drinks because it is made neither from fruit nor from grain. Technically, it is closer to wine than to beer because the fermentation occurs directly from sugars, as in wine (usually from fruit), as opposed to from starches (usually from grain) converted to sugars by mashing, as in beer. But in terms of experience and traditional manner of consumption, it is much more comparable to beer. It is even milder in alcoholic content than beer. It is arguably the region's beer equivalent.

Kumis is very light in body compared to most dairy drinks. It has a unique, slightly sour flavor with a bite from the mild alcoholic content. The exact flavor is greatly variable between different producers.

Kumis is usually served cold or chilled. Traditionally it is sipped out of small, handle-less, bowl-shaped cups or saucers, called piyala. The serving of it is an essential part of Kyrgyz hospitality on the jayloo or high pasture, where they keep their herds of animals (horse, cattle, and sheep) during the summer phase of transhumance.

VIILI

Viili or filbunke is a mesophilic fermented milk product found in Finland that originated in Scandinavia. This cultured milk beverage is the results of microbial action of lactic acid bacteria (LAB) and a surface-growing yeast-like fungus Geotrichum candidum present in milk, which forms a velvet-like surface on viili. In addition, most traditional viili cultures also contain yeast strains such as Kluveromyces marxianus and Pichia fermentans. The LAB identified in viili including Lactococcus lactis subsp. cremoris, Lactococcus lactis subsp. lactis biovar. diacetylactis, Leuconostoc mesenteroides subsp. cremoris. Among those mesophilic LAB strains, the slime-forming Lc. lactis subsp. cremoris produce a phosphate-containing heteropolysaccharide, named viilian. Viilian is similar to kefiran produced by kefir grains. The production of exopolysaccharides (EPS) by the strain forms the consistency character of viili and it has been claimed to have various functional benefits toward the rheological properties of milk products and the health improving potential.

In modern practice, pasteurized milk is used, fermentation is carried out in a dairy plant in controlled conditions using laboratory-grown cultures and the product sold fresh. Viili is widely available in Finland in grocery stores in several variants.

Other Variants

Several variants of fermented milk products are found in Western Finland and Sweden, such as filmjölk ("viili milk") or långfil ("long viili"), which vary in consistency and fermentation. In Norway, filmjölk is usually named "kulturmelk" ("cultured milk") or "surmelk" ("sour milk"), while in Gotland and Iceland, the name "skyr" is used to refer to fermented yoghurt variants.

Cream viili is made from cream instead of milk, and is used in cooking like sour cream, or with dill, chives and other spices as cold sauce for fish, or as a base for dip sauces.

BUTTERMILK

Buttermilk is a fermented dairy drink. Traditionally, it was the liquid left behind after churning butter out of cultured cream; most modern buttermilk is cultured, however. It is common in warm climates (including the Balkans, South Asia, the Middle East and the Southern United States) where unrefrigerated fresh milk sours quickly.

Buttermilk can be drunk straight, and it can also be used in cooking. In making soda bread, the acid in buttermilk reacts with the raising agent, sodium bicarbonate, to produce carbon dioxide which acts as the leavening agent. Buttermilk is also used in marination, especially of chicken and pork, which the lactic acid helps to tenderize, retain moisture and allows added flavors to permeate the meat.

Traditional Buttermilk

Originally, buttermilk referred to the liquid left over from churning butter from cultured or fermented cream. Traditionally, before the advent of homogenization, the milk was left to sit for a period of time to allow the cream and milk to separate. During this time, naturally occurring lactic acid-producing bacteria in the milk fermented it. This facilitates the butter churning process, since fat from cream with a lower pH coalesces more readily than that of fresh cream. The acidic environment also helps prevent potentially harmful microorganisms from growing, increasing shelf-life.

Traditional buttermilk is still common in many Indian, Nepalese, and Pakistani households, but rarely found in Western countries. In Nepal, buttermilk is called *mohi* and is a common drink in many Nepalese homes. It is served to family members and guests, and can be taken with meals or snacks. In many families, it is most popularly served with roasted maize.

Cultured Buttermilk

Cultured buttermilk was first commercially introduced in the United States in the 1920s. Commercially available cultured buttermilk is milk that has been pasteurized and homogenized, and then inoculated with a culture of Lactococcus lactis or Lactobacillus bulgaricus plus Leuconostoc citrovorum to simulate the naturally occurring bacteria in the old-fashioned product. The tartness of cultured buttermilk is primarily due to lactic acid produced by lactic acid bacteria while fermenting lactose, the primary sugar in milk. As the bacteria produce lactic acid, the pH of the milk decreases and casein, the primary milk protein, precipitates, causing the curdling or clabbering of milk. This process makes buttermilk thicker than plain milk. While both traditional and cultured buttermilk contain lactic acid, traditional buttermilk tends to be less viscous, whereas cultured buttermilk is more viscous.

When introduced, cultured buttermilk was popular among immigrants, and viewed as a food that could slow aging. It reached peak annual sales of 517,000,000 kilograms (1.140×10^9 lb) in 1960. Buttermilk's popularity has declined since then, despite an increasing population, and annual sales in 2012 reached less than half that number.

However, condensed buttermilk and dried buttermilk remain important in the food industry. Liquid buttermilk is used primarily in the commercial preparation of baked goods and cheese. Buttermilk solids are used in ice cream manufacturing, as well as being added to pancake mixes to make buttermilk pancakes.

Acidified Buttermilk

Acidified buttermilk is a substitute made by adding a food-grade acid such as vinegar or lemon juice to milk. It can be produced by mixing 1 tablespoon (0.5 US fluid ounces, 15 ml) of acid with 1 cup (8 US fluid ounces, 240 ml) of milk and letting it sit until it curdles, about 10 minutes. Any level of fat content for the milk ingredient may be used, but whole milk is usually used for baking. In the process which is used to produce paneer, such acidification is done in the presence of heat.

Nutrition

Buttermilk is comparable to regular milk in terms of calories and fat. One cup (237 ml) of whole milk contains 157 calories and 8.9 grams of fat. One cup of whole buttermilk contains 152 calories and 8.1 grams of total fat. Low fat buttermilk is also available. Buttermilk contains vitamins, potassium, calcium, and traces of phosphorus.

Chemical Composition

The chemical composition of buttermilk varies to a great extent, depending on the amount of water added to cream. Some of the butter manufacturers standardize cream with water, thereby decreasing the total solids level of buttermilk. The gross chemical composition of buttermilk produced under ideal conditions is almost similar to that of skim milk.

Characteristic	Skim Milk	Sweet Cream Buttermilk
T.S. (%)	10.38	9.88
Fat (%)	0.09	0.59
Total proteins (%)	4.27	3.73
Lactose (%)	5.2	4.81
Ash (%)	0.82	0.75
Total phospholipids (mg %)	8.65	78.56

Titratable acidity (% LA)	0.16	0.12
PH	6.69	6.86
Curd tension (g)	66.85	18.84
Relative viscosity (cP at 300 C)	1.64	1.80

Average gross composition and physico-chemical properties of sweet cream buttermilk and skim milk (obtained from buffalo milk).

Sour buttermilk differs from sweet cream buttermilk in respect of titratable acidity.The acidity in sweet cream buttermilk varies from 0.10 to 0.14 per cent, whereas in sour buttermilk it is even as high as 1%. However, there is not much difference in the chemical composition of two types of buttermilk. buttermilk has wide range of composition depending on the quality of milk used for making curd and levels of addition of water during churning. buttermilk, on an average, contains 4% total solids comprising of 0.8% fat, 1.29% protein and 1.2% lactic acidity. The colour of buttermilk is brownish due to prolonged heating of milk before culturing and the body not as homogeneous as that of factory produced buttermilk. When kept undisturbed for sometime, curdy material deposits at the bottom of buttermilk.

Processing and Drying of Sweet Cream Buttermilk

Being almost similar in gross chemical composition of skim milk, no problem is encountered during its processing, i.e., separation, clarification, pasteurization, concentration and drying. Rather the heat stability of sweet cream buttermilk is considered to be better than skim milk thereby making it more suitable for processing to very high heat treatments. Concentration and spray drying of sweet cream buttermilk can also be achieved adopting the same standard conditions used for skim milk. The physico-chemical properties of spray dried sweet cream buttermilk and skim milk.

Characteristic	Skim milk powder	Sweet cream buttermilk powder
Moisture (%)	2.75	2.59
Fat (%)	1.05	6.38
Total protein (%)	40.29	37.09
Lactose (%)	48.15	47.00
Ash (%)	7.76	6.94
Total phospholipids (mg %)	97.1	625.25
Titratable acidity (% L.A)	1.39	1.17
Solubility index (ml)	0.30	0.15
Bulk density (g/ml)	0.544	0.345

Physico-chemical Characteristics of Spray Powders

The striking differences between two types of powders are the high total lipids including phospholipids and low bulk density in sweet cream buttermilk powder in comparison with skim milk. The spray dried buttermilk powder is less free flowing and dusty because of high fat content in comparison with skim milk powder. Though the high fat content reduces the shelf life of the powder during storage, the high phospholipids will provide better oxidative stability to dried buttermilk.

Utilisation of Sweet Cream Buttermilk

Sweet cream buttermilk, because of its resemblance in gross chemical composition with skim milk, is usually admixed with bulk of skim milk for further spray drying or even product manufacture in dairy plants. Sweet cream buttermilk can be used in beverage form and in the fluid milk industry as a milk extender with specific benefits over skim milk. The other potential uses of buttermilk solids are in manufacture of soft varieties of cheese, paneer, fermented milks and traditional milk products.

However, various physico-chemical properties of buttermilk differ from that of skim milk. Sweet cream buttermilk has lower acidity and curd tension but higher viscosity as compared with skim milk. These differences in physico-chemical properties of buttermilk and skim milk provide many choices for their selective applications in dairy products manufacture. Buttermilk contains higher fat content than skim milk, which can be reduced to some extent by subjecting it to centrifugal separation. Buttermilk contains a larger proportion of protein mixture sloughed from the fat globule-milk-serum interface by churning process. The amount of fat globule membrane protein (FGMP) is, however, not as large in comparison with total buttermilk proteins. The FGMP are hydrophilic and hydrophobic in nature and their physical properties, nitrogen content and amino acid composition do not correspond with any other milk proteins. The FGMP also contributes a complex mixture of glycerophospholipids to buttermilk. Sweet cream buttermilk contains about nine times higher phospholipids than skim milk. It has been noticed that phospholipids in buttermilk do not have short chain fatty acids. The principal fatty acids are C16 (palmitic) and higher acids. Of the total phospholipid fatty acids, about 40% by wt. are saturated acids and the rest are non-conjugated di- to pentaunsaturated acids. Phospholipids of buttermilk include more or less equal proportions of lecithin, sphingomyelin and cephalin together with a small proportion of cerebrosides.

- Beverage: As beverage, buttermilk is consumed in plain and spiced forms throughout the year and highly used as a refreshing drink in summer season. A number of state federations and private plants sell plain buttermilk in 500 ml and 1 kg pack and salted and spiced buttermilk in 200 ml pouches. "Sumul chhach" is packed in 500 ml packs.

- Market milk: The undiluted sweet cream buttermilk produced in the organized dairies is partly admixed with the whole milk for fluid milk supply. It has been observed that use of sweet cream buttermilk in the market milk for toning of buffalo milk improves the palatability, viscosity and heat-stability and reduce the curd tension without adversely affecting the keeping quality. In addition to plain fluid milk, it can also be used for the preparation of flavoured milks and milk beverages. The powder made from the mixture of skim milk and sweet cream buttermilk is treated as a skim milk powder and used for reconstitution purposes.

- Fermented milk product: Curd prepared by incorporating sweet cream buttermilk into whole milk has soft-body which is probably due to the change in the electric charge on the casein during churning, the presence of phospholipids and other FGM materials, and the free fat in the buttermilk. Addition of 1-2% skim milk powder is recommended for improving the body of dahi made from buttermilk. As an alternative to curd making, sweet cream buttermilk can be successfully utilized in the manufacture of cultured buttermilk and lassi in which the firmness is not of much consideration.

- Paneer: Buffalo milk has to be standardized to a fat and SNF ratio of about 1:1.65 to meet the PFA requirements for the manufacture of paneer. The replacement of skim milk with sweet cream buttermilk for the standardization of buffalo milk has been found to increase the yield of paneer by about one per cent without altering the organoleptic and textural properties. It is also possible to prepare good quality paneer from low fat milk by incorporating buttermilk solids to buffalo milk.

- Cheese: The preparation of hard varieties of cheese like Cheddar and Gouda involves the adjustment of casein and fat ratio with the help of skim milk. The replacement of skim milk with sweet cream buttermilk results into softer body due to the presence of higher amount of fat globule membrane materials in buttermilk. Several benefits of utilizing buttermilk solids in the manufacture of soft varieties of cheeses are: decreased waste disposal problems at the creamery, reduction in cost, increased cheese yield and improved flavour, texture, biological value and hypocholesterolaemic effects of cheese.

- Other uses: Sweet cream buttermilk can also be used for manufacture of some popular indigenous dairy products. The buttermilk powder can also be used in the preparation of ice cream and bakery products.

References

- Dairy-product: britannica.com, Retrieved 21 May, 2019

- Chandan, Ramesh C.; Kilara, Arun (22 December 2010). Dairy Ingredients for Food Processing. John Wiley & Sons. Pp. 1–. ISBN 978-0-470-95912-1

- General Mills to discontinue producing Colombo Yogurt". Eagle-Tribune. 29 January 2010. Archived from the original on 28 May 2011. Retrieved 29 April 2010

- Connaitre-creme-industrielle: fitsa-group.com, Retrieved 26 July, 2019

- "What is sour cream. Sour cream for cooking recipes". Homecooking.about.com. 2010-06-14. Retrieved 2011-09-14

- Hui, Y. H (2004-01-01). Handbook of food and beverage fermentation technology. New York: Marcel Dekker. ISBN 978-0824751227

- Katragadda, H. R.; Fullana, A. S.; Sidhu, S.; Carbonell-Barrachina, Á. A. (2010). "Emissions of volatile aldehydes from heated cooking oils". Food Chemistry. 120: 59. Doi:10.1016/j.foodchem.2009.09.070

- Whipped-cream-structure, foodscience: uoguelph.ca, Retrieved 5 February, 2019

- The Culinary Institute of America (2011). The Professional Chef (9th ed.). Hoboken, New Jersey: John Wiley & Sons. ISBN 978-0-470-42135-2. OCLC 707248142

- Clabber-milk, traditional-cooking-traditional-living: butterforall.com, Retrieved 19 April, 2019

- 9-health-benefits-of-kefir, nutrition: healthline.com, Retrieved 17 May, 2019

- Buttermilk: dairy-technology.blogspot.com, Retrieved 14 July, 2019

Dairy Processing

Dairy processing includes various processes and methods such as thermization, milk powder production, clarification and cream separation, sterilization, homogenization, fluid milk production, pasteurization, etc. This chapter closely examines these techniques of dairy processing to provide an extensive understanding of the subject.

The Milk Processing section contains general information on operations important in milk processing.

From the Farm to the Processing Plant

To provide the safest and highest quality product to the consumer, the Pasteurized Milk Ordinance provides standardized guidelines. Each state regulates their own dairy industry, but the state's guidelines usually meet or exceed those defined by the PMO. Milk that is shipped between states must follow the PMO regulations.

Milk is obtained from the cow (or goat, sheep, or water buffalo) under sanitary conditions and cooled to 45 °F (7 °C) within 2 hours of milking. Milk is picked up by a handler who takes a sample and then pumps the milk from farm's bulk tank into the milk truck. A handler may pick up milk from more than one farm, so a truck load may contain milk from several farms when it is delivered to the processing plant. Before the milk can be unloaded at the processing plant, each load is tested for antibiotic residues. If the milk shows no evidence of antibiotics, it is pumped into the plant's holding tanks for further processing. If the milk does not pass antibiotic testing, the entire truck load of milk is discarded and the farm samples are tested to find the source of the antibiotic residues. Regulatory action is taken against the farm with the positive antibiotic test. Positive antibiotic tests are rare, and account for far less than 1% of the tank loads of milk delivered to processing plants.

Milk at the plant is stored at less than 45 °F (7 °C) and is usually processed within 24 hours, but can held for up to 72 hours (3 days) before processing. Longer holding time allows for growth of spoilage organisms that grow at refrigerator temperatures, called psychrotrophs. Almost all of the milk in the United States is pasteurized, with the exception of some raw milk cheese production and some states that allow the sale of raw milk. The conditions of the heat treatment used for pasteurization depends on the final product - lower temperatures are used for refrigerated products and higher heat treatments are used for products stored at room temperature.

THERMIZATION

Thermization, also spelled thermisation, is a method of sanitizing raw milk with low heat. "Thermization is a generic description of a range of subpasteurization heat treatments (57 to 68 °C × 10 to 20 s) that markedly reduce the number of spoilage bacteria in milk with minimal heat damage." The process is not used on other food products, and is similar to pasteurization but uses lower temperatures, allowing the milk product to retain more of its original taste. In Europe, there is a distinction between cheeses made of thermized milk and raw-milk cheeses. However, the United States' Food and Drug Administration (FDA) places the same regulations on all unpasteurized cheeses. As a result, cheeses from thermized milk must be aged for 60 days or more before being sold in the United States, the same restriction placed on raw-milk cheeses by the FDA.

Some cheeses, including varieties of blue cheese, are made from thermized milk.

Thermization involves heating milk at temperatures of around 145–149 °F (63–65 °C) for 15 seconds, while pasteurization involves heating milk at 160 °F (71 °C) for 15 seconds or at 145 °F (63 °C) for 30 minutes. Thermization is used to extend the keeping quality of raw milk (the length of time that milk is suitable for consumption) when it cannot be immediately used in other products, such as cheese. Thermization can also be used to extend the storage life of fermented milk products by inactivating microorganisms in the product.

Thermization inactivates psychrotrophic bacteria in milk and allows the milk to be stored below 8 °C (46 °F) for three days, or stored at 0–1 °C (32–34 °F) for seven days. Later, the milk may be given stronger heat treatment to be preserved longer. Cooling thermized milk before reheating is necessary to delay/prevent the outgrowth of bacterial spores. When the milk is first heated, spores can begin to germinate, but their growth can be halted or delayed when the milk is refrigerated, depending on the microorganisms' growth requirements. Germinated spores are sensitive to subsequent heating, however since germination is not a homogeneous process, not all spores will germinate or be inactivated by subsequent heating.

CLARIFICATION AND CREAM SEPARATION

Centrifugal separation is a process used quite often in the dairy industry. Some uses include:

- Clarification (removal of solid impurities from milk prior to pasteurization).

- Skimming (separation of cream from skim milk).

- Standardizing.

- Whey separation (separation of whey cream (fat) from whey).

- Bactofuge treatment (separation of bacteria from milk).

- Quark separation (separation of quarg curd from whey).

- Butter oil purification (separation of serum phase from anhydrous milk fat).

Principles of Centrifugation

Centrifugation is based on Stoke's Law. The particle sedimentation velocity increases with:

- Increasing diameter.

- Increasing difference in density between the two phases.

- Decreasing viscosity of the continuous phase.

If raw milk were allowed to stand, the fat globules would begin to rise to the surface in a phenomena called creaming. Raw milk in a rotating container also has centrifugal forces acting on it. This allows rapid separation of milk fat from the skim milk portion and removal of solid impurities from the milk.

Clarification

Separation and clarification can be done at the same time in one centrifuge. Particles, which are more dense than the continuous milk phase, are thrown back to the perimeter. The solids that collect in the centrifuge consist of dirt, epithelial cells, leucocytes, corpuscles, bacteria sediment and sludge. The amount of solids that collect will vary, however, it must be removed from the centrifuge.

The following image is a schematic of both a clarifier and a separator:

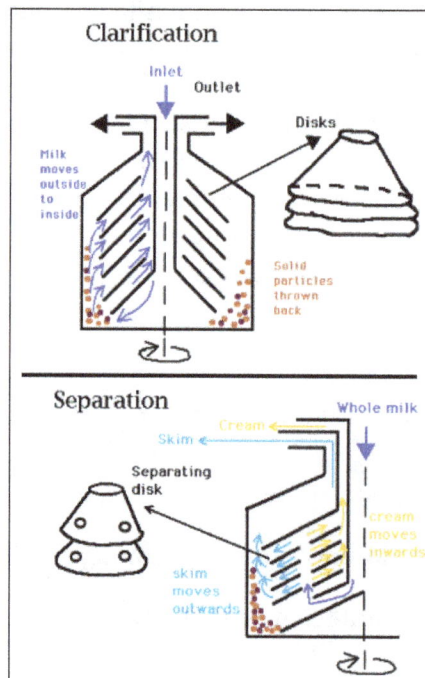

More modern centrifuges are self-cleaning allowing a continuous separation/clarification process. This type of centrifuge consists of a specially constructed bowl with peripheral discharge slots. These slots are kept closed under pressure. With a momentary release of pressure, for about 0.15 s, the contents of sediment space are evacuated. This can mean anywhere from 8 to 25 L are ejected at intervals of 60 min. For one dairy, self-cleaning translated to a loss of 50 L/hr of milk.

Separation

Centrifuges can be used to separate the cream from the skim milk. The centrifuge consists of up to 120 discs stacked together at a 45 to 60 degree angle and separated by a 0.4 to 2.0 mm gap or separation channel. Milk is introduced towards the inner edge of the disc stack. The stack of discs has vertically aligned distribution holes into which the milk is introduced.

Under the influence of centrifugal force the fat globules, which are less dense than the skim milk, move inwards through the separation channels toward the axis of rotation. Some skim is needed to carry the fat globules out of the separation, and the combination of fat globules in a much-reduced volume of skim milk is called "cream", i.e., creamis skim milk enriched in fat globules. The skim milk, now devoid of fat globules, will move outwards and leaves through a separate outlet.

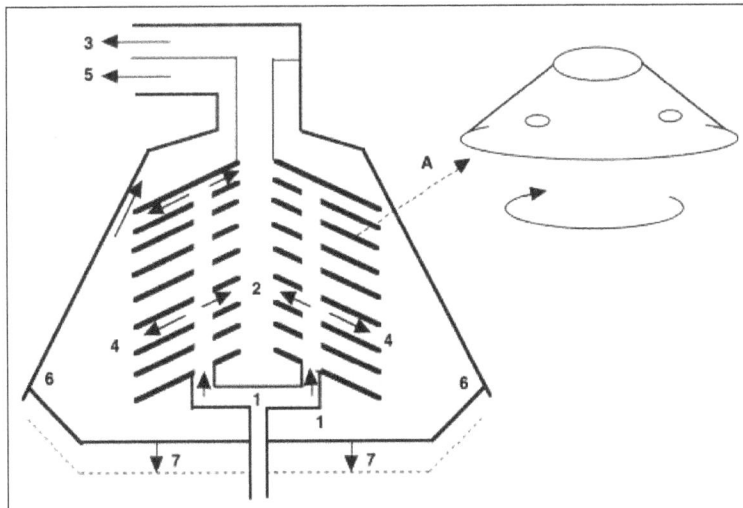

A Conical disk, from stack, showing top opening and milk distribution channels, Disk stack is rotating in separator:

1. Whole milk enters the disk distribution channel.

2. Fat globules move inward, carried by some skim milk to from cream.

3. Cream outlet port.

4. Skim milk devoid of fat globules moves outward and upward, over the solid disk at the top.

5. Skim outlet port.

6. Sediment collects.

7. Momentary periodic release of bowl pressure causes sediment cavity (6) to open, releasing sediment.

Standardization

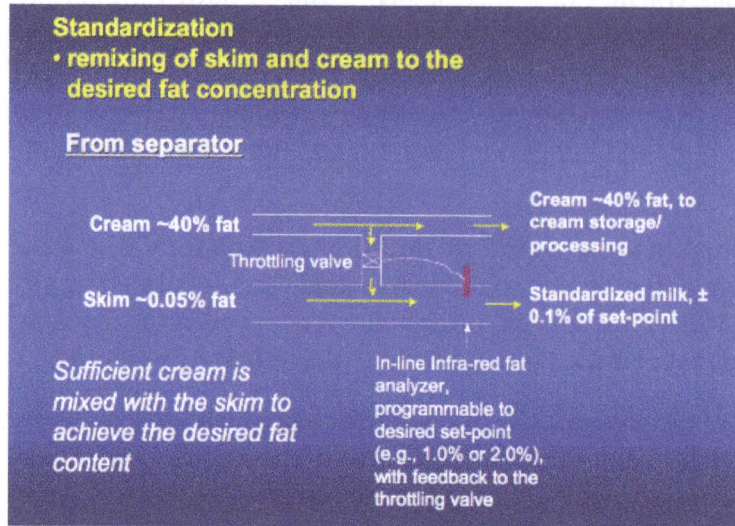

The streams of skim and cream after separation must be recombined to a specified fat content. This can be done by adjusting the throttling valve of the cream outlet; if the valve is completely closed, all milk will be discharged through the skim milk outlet. If the valve is very slightly opened, you get most of the fat globules in a very small volume of skim, so cream with a high fat content. As the valve is progressively opened further, larger volumes of skim push the fat globules out, so you get larger volumes of cream but with diminishing fat contents (%) being discharged from the cream outlet. With direct standardization the cream and skim are automatically remixed at the separator to provide the desired fat content.

Standardization Problems

Here are the two examples of how to do milk standardization calculations:

If a dairy has 160 kg of 40% fat cream and wishes to standardize it to 32% fat cream, how much skimmilk (0% fat) must be added?

Mass Balance Approach:

let x = kg skimmilk

y = kg of 32% cream

Mass Balance Equation (total mass into the process = total mass out of the process) 160 + x = y

Component Balance for Fat (fat into the process = fat out of the process)

.40 (160) = .32 (y) which says 40% of 160 kg comes in and 32% of y goes out

.40 (160) = .32(160+ x), substituting our equation for y

.32x = 64 - 51.2

x = 40 kg skimmilk

How much cream testing 35% fat must be added to 500 kg of milk testing 4% fat to obtain cream testing 10% fat?

> let x = kg 35% cream

> y = kg of 10% cream

Mass Balance x + 500 = y (mass in = mass out)

Component Balance for Fat:

> x (.35) + 500 (.04) = y (.10) again, fat in = fat out

> x (.35) + 20 = x(.10) + 50

> .25 x = 30

> x = 120 kg of 35% cream

Pearson Square Approach to above Problem

- A shortcut for 2-component mass balances.
- Place desired percentage in the centre of a rectangle.
- Place percentage composition of two available streams in left corners of rectangle.
- Cross subtract lower from higher numbers for right corners of triangle (higher number - lower number).
- Use right corners as ratios of two streams.

Component Mass:

```
35      6
    10
 4    25    500
      ──  ───
      31
```

35% cream = (500 x 6)/25 = 120 kg

This will produce 620 kg cream at 10% fat.

Check yur work - (620 x 0.10) = (120 x .35) + (500 x .04) - correct.

Thermal Destruction of Microorganisms

Heat is lethal to microorganisms, but each species has its own particular heat tolerance. During a thermal destruction process, such as pasteurization, the rate of destruction is logarithmic, as is their rate of growth. Thus bacteria subjected to heat are killed at a rate that is proportional to the number of organisms present. The process is dependent both on the temperature of exposure and the time required at this temperature to accomplish to desired rate of destruction. Thermal calculations thus involve the need for knowledge of the concentration of microorganisms to be destroyed, the acceptable concentration of microorganisms that can remain behind (spoilage organisms, for example, but not pathogens), the thermal resistance of the target microorganisms (the most heat tolerant ones), and the temperature time relationship required for destruction of the target organisms.

The extent of the pasteurization treatment required is determined by the heat resistance of the most heat-resistant enzyme or microorganism in the food. For example, milk pasteurization historically was based on Mycobacterium tuberculosis and Coxiella burnetti, but with the recognition of each new pathogen, the required time temperature relationships are continuously being examined.

A thermal death curve for this process is shown below. It is a logarithmic process, meaning that in a given time interval and at a given temperature, the same percentage of the bacterial population will be destroyed regardless of the population present. For example, if the time required to destroy one log cycle or 90% is known, and the desired thermal reduction has been decided (for example, 12 log cycles), then the time required can be calculated. If the number of microorganisms in the food increases, the heating time required to process the product will also be increased to bring the population down to an acceptable level. The heat process for pasteurization is usually based on a 12 D concept, or a 12 log cycle reduction in the numbers of this organism.

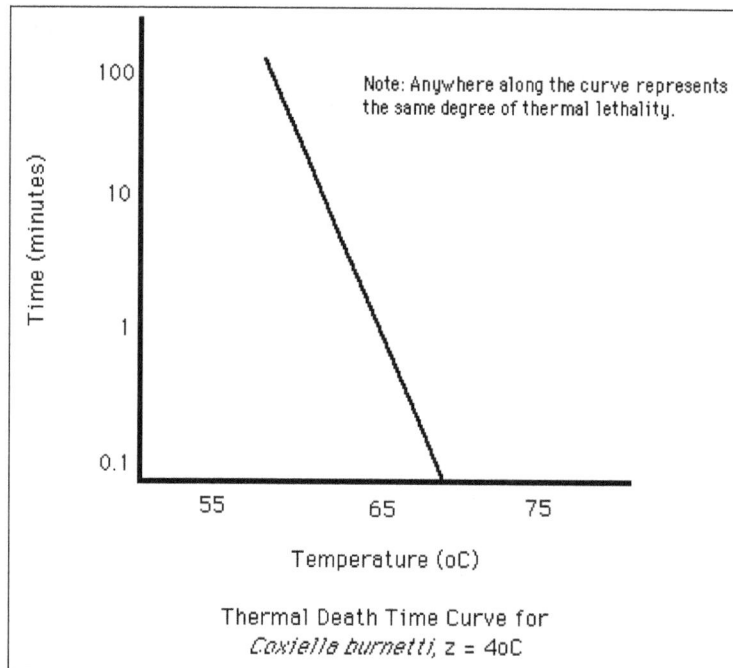

Note: Anywhere along the curve represents the same degree of thermal lethality.

Thermal Death Time Curve for
Coxiella burnetti, z = 4oC

Several parameters help us to do thermal calculations and define the rate of thermal lethality. The D value is a measure of the heat resistance of a microorganism. It is the time in minutes at a given temperature required to destroy 1 log cycle (90%) of the target microorganism. For example, a D value at 72 °C of 1 minute means that for each minute of processing at 72 °C the bacteria population of the target microorganism will be reduced by 90%. In the illustration below, the D value is 14 minutes (40-26) and would be representative of a process at 72 °C.

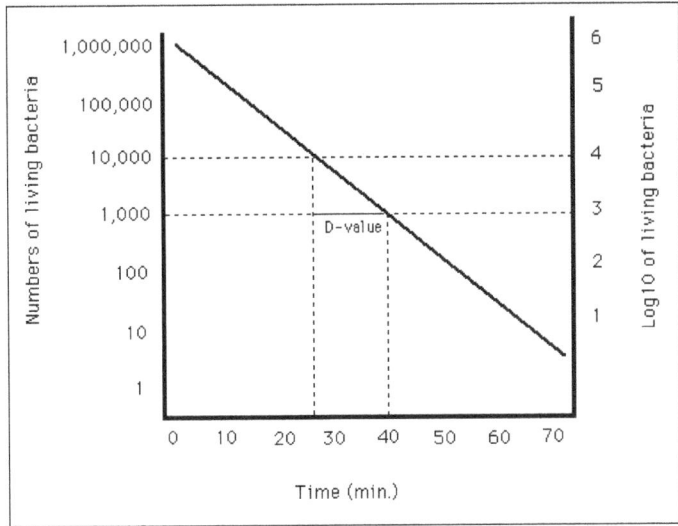

The Z value reflects the temperature dependence of the reaction. It is defined as the temperature change required to change the D value by a factor of 10. In the illustration below the Z value is 10 °C.

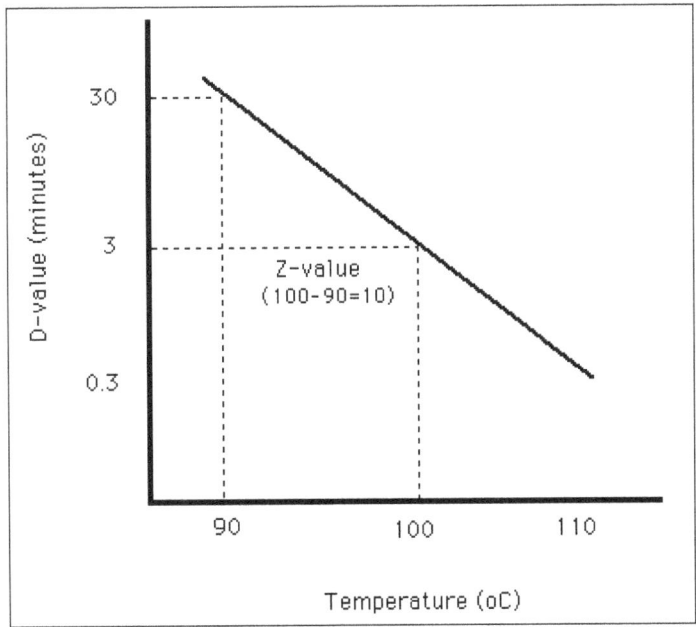

Reactions that have small Z values are highly temperature dependent, whereas those with large Z values require larger changes in temperature to reduce the time. A Z value of 10 °C is typical for a spore forming bacterium. Heat induced chemical changes have much larger Z values that micro-organisms, as shown below.

- Bacteria Z (°C) 5-10 D121 (min) 1-5.

- Enzymes Z (°C) 30-40 D121 (min) 1-5.

- Vitamins Z (°C) 20-25 D121 (min) 150-200.

- Pigments Z (°C) 40-70 D121 (min) 15-50.

The figure below (which is schematic and not to scale) illustrates the relative changes in time temperature profiles for the destruction of microorganisms. Above and to the right of each line the microorganisms or quality factors would be destroyed, whereas below and to the left of each line, the microorganisms or quality factors would not be destroyed. Due to the differences in Z values, it is apparent that at higher temperatures for shorter times, a region exists (shaded area) where pathogens can be destroyed while vitamins can be maintained. The same holds true for other quality factors such as colour and flavour components. Thus in UHT milk processing, very high temperatures for very short times (e.g., 140 °C for 1-2 s) are favoured compared to a lower temperature longer time processes since it results in bacterial spore elimination with a lower loss of vitamins and better sensory quality.

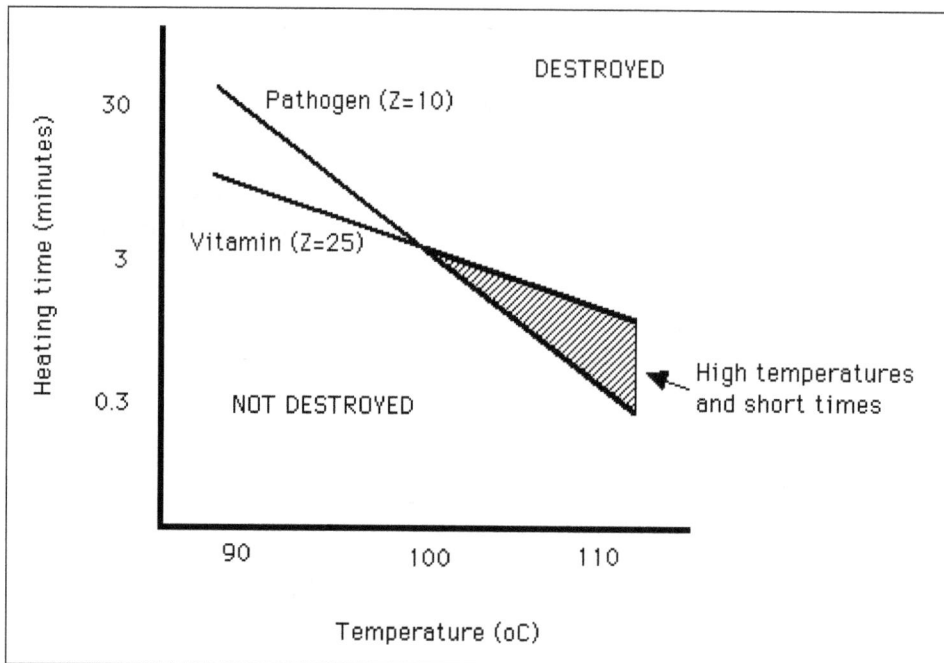

Alkaline phosphatase is a naturally-occurring enzyme in raw milk which has a similar Z value to heat-resistant pathogens. Since the direct estimation of pathogen numbers by microbial methods is expensive and time consuming, a simple test for phosphatase activity is routinely used. If activity is found, it is assumed that either the heat treatment was inadequate or that unpasteurized milk has contaminated the pasteurized product.

A working example of how to use D and Z values in pasteurization calculations:

Pooled raw milk at the processing plant has bacterial population of 4x10exp5/mL. It is to be processed at 79 °C for 21 seconds. The average D value at 65 °C for the mixed population is 7 min. The Z value is 7 °C. How many organisms will be left after pasteurization? What time would be required at 65 °C to accomplish the same degree of lethality?

At 79 °C, the D value has been reduced by two log cycles from that at 65 °C since the Z value is 7 °C. Hence it is now 0.07 min. The milk is processed for 21/60=0.35 min, so that would accomplish 5 log cycle reductions to 4 organisms/mL. At 65 °C, you would need 35 minutes to accomplish a 5D reduction.

MILK POWDER PRODUCTION

Milk powder manufacture is a simple process now carried out on a large scale. It involves the gentle removal of water at the lowest possible cost under stringent hygiene conditions while retaining all the desirable natural properties of the milk - colour, flavour, solubility, nutritional value. Whole (full cream) milk contains, typically, about 87% water and skim milk contains about 91% water. During milk powder manufacture, this water is removed by boiling the milk under reduced pressure at low temperature in a process known as evaporation. The resulting concentrated milk is then sprayed in a fine mist into hot air to remove further moisture and so give a powder. Approximately 13 kg of whole milk powder (WMP) or 9 kg of skim milk powder (SMP) can be made from 100 L of whole milk.

Milk powders may vary in their gross composition (milkfat, protein, lactose), the heat treatment they receive during manufacture, powder particle size and packaging. Special "high heat" or "heat-stable" milk powders are required for the manufacture of certain products such as recombined evaporated milk. Milk powders of various types are used in a wide range of products such as baked goods, snacks and soups, chocolates and confectionary (e.g. milk chocolate), ice cream, infant formulae, nutritional products for invalids, athletes, hospital use etc., recombined milks and other liquid beverages.

Marco Polo in the 13th century reported that soldiers of Kublai Khan carried sun-dried milk on their expeditions. In more recent times, milk has been dried in thin films on heated rollers. The earliest patents for this process date from the turn of the century. Such roller drying was the main means of producing milk powders until the 1960s when spray drying took over. Milk powder manufacture is now very big business.

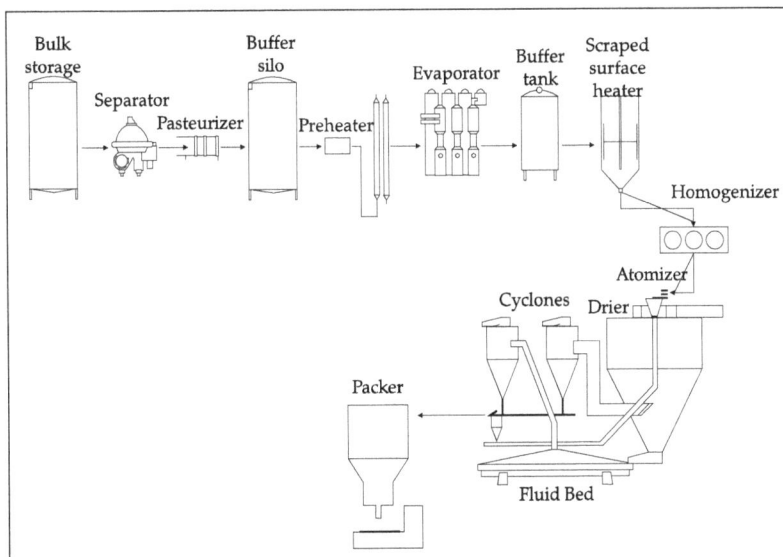

Milk powder manufacture is a simple process now carried out on a large scale. It involves the gentle removal of water at the lowest possible cost under stringent hygiene conditions while retaining all the desirable natural properties of the milk - colour, flavour, solubility, nutritional value. Whole (full cream) milk contains, typically, about 87% water and skim milk contains about 91% water.

During milk powder manufacture this water is removed by boiling the milk under reduced pressure at low temperature in a process known as evaporation. The resulting concentrated milk is then sprayed in a fine mist into hot air to remove further moisture and so give a powder. Approximately 13 kg of whole milk powder (WMP) or 9 kg of skim milk powder (SMP) can be made from 100 L of whole milk. The milk powder manufacturing process is shown in the following schematic and is described in detail above.

Separation/Standardization

The conventional process for the production of milk powders starts with taking the raw milk received at the dairy factory and pasteurising and separating it into skim milk and cream using a centrifugal cream separator. If WMP is to be manufactured, a portion of the cream is added back to the skim milk to produce a milk with a standardised fat content (typically 26- 30% fat in the powder). Surplus cream is used to make butter or anhydrous milkfat.

Preheating

The next step in the process is "preheating" during which the standardised milk is heated to temperatures between 75 and 12 °C and held for a specified time from a few seconds up to several minutes (pasteurisation: 72 °C for 15 s). Preheating causes a controlled denaturation of the whey proteins in the milk and it destroys bacteria, inactivates enzymes, generates natural antioxidants and imparts heat stability. The exact heating/holding regime depends on the type of product and its intended end-use. High preheats in WMP are associated with improved keeping quality but reduced solubility. Preheating may be either indirect (via heat exchangers), or direct (via steam injection or infusion into the product), or a mixture of the two. Indirect heaters generally use waste heat from other parts of the process as an energy saving measure.

Evaporation

In the evaporator the preheated milk is concentrated in stages or "effects" from around 9.0% total solids content for skim milk and 13% for whole milk, up to 45-52% total solids. This is achieved by boiling the milk under a vacuum at temperatures below 72 °C in a falling film on the inside of vertical tubes, and removing the water as vapour. This vapour, which may be mechanically or thermally compressed, is then used to heat the milk in the next effect of the evaporator which may be operated at a lower pressure and temperature than the preceding effect. Modern plants may have up to seven effects for maximum energy efficiency. More than 85% of the water in the milk may be removed in the evaporator. Evaporators are extremely noisy because of the large quantity of water vapour travelling at very high speeds inside the tubes.

Spray Drying

Spray drying involves atomising the milk concentrate from the evaporator into fine droplets. This is done inside a large drying chamber in a flow of hot air (up to 20 °C) using either a spinning disk atomiser or a series of high pressure nozzles. The milk droplets are cooled by evaporation and they never reach the temperature of the air. The concentrate may be heated prior to atomisation to reduce its viscosity and to increase the energy available for drying. Much of the remaining water is evaporated in the drying chamber, leaving a fine powder of around 6% moisture content with a mean particle size typically of < 0.1 mm diameter. Final or "secondary" drying takes place in a fluid bed, or in a series of such beds, in which hot air is blown through a layer of fluidised powder removing water to give product with a moisture content of 2-4%. Precautions must be taken to prevent fires and to vent dust explosions should they occur in the drying chamber or elsewhere. Such explosions can be extremely dangerous to life, property and markets.

Packaging and Storage

Milk powders are immensely more stable than fresh milk but protection from moisture, oxygen, light and heat is needed in order to maintain their quality and shelf life. Milk powders readily take up moisture from the air, leading to a rapid loss of quality and caking or lumping. The fat in WMPs can react with oxygen in the air to give off-flavours, especially at higher storage temperatures (> 3 °C) typical of the tropics.

Milk powder is packed into either plastic-lined multi-wall bags (25 kg) or bulk bins (600 kg). WMPs are often packed under nitrogen gas to protect the product from oxidation and to maintain their flavour and extend their keeping quality. Packaging is chosen to provide a barrier to moisture, oxygen and light. Bags generally consist of several layers to provide strength and the necessary barrier properties. Shipments of milk powder should never suffer prolonged exposure to direct sunshine especially in tropical countries. A few hours at elevated temperatures (> 4 °C) during transshipment can negate many weeks of careful storage.

Agglomerated Powders

Standard powders, because of their fine dusty nature, do not reconstitute well in water. "Agglomerated" and "instant" powders were specifically developed to counter this. The manufacture of an agglomerated powder initially follows the standard process of evaporation and drying, described above. However, during spray drying small particles of powder leaving the drier (the "fines") are recovered in cyclones and returned to the drying chamber in the close proximity of the atomiser. The wet concentrate droplets collide with the fines and stick together, forming larger (0.1-0.3 mm), irregular shaped "agglomerates". Agglomerated powders disperse in water more rapidly and are less dusty and easier to handle than standard powders.

Instant whole Milk Powder

With WMP, an extra step is required after agglomeration to make the product truly "instant" and overcome the hydrophobic (water-hating) nature of traces of free fat on the surface of the particles. This extra step consists of spraying minute quantities of the natural surfactant or wetting agent, soy lecithin, on to the powder in a fluid bed. Soy lecithin is extracted from soy bean oil. Lecithins are widespread in nature and they occur naturally in milk.

Energy and Environmental Considerations

Large amounts of energy are expended in the process of removing water and so plants developed over the years have become increasingly more energy efficient. Evaporators are much more energy efficient than driers, using only a fraction of a kilogram of steam (or the energy equivalent) per kilogram of water removed. Driers on the other hand use several kilograms of steam (or steam equivalent) per kilogram of water evaporated. Spray drying provides a means of rapidly and gently removing the bulk of the remaining water but, ideally, spray driers have short residence times. Hence fluid beds are used for the final stages of drying. The powder remains for several minutes in fluid beds allowing time for the last of the water to be removed.

Milk powder manufacturing plants tend to be very large, few in number and located in rural areas. If modern and well managed, they have only relatively small effects on the environment. They are moderately energy intensive, burning coal or gas and consuming substantial electricity. There are strong economic pressures to reduce energy consumption but there is little scope for further major improvement. Milk storage silos, cream separators and the evaporators and associated plant must be cleaned daily, and driers less often. Sodium hydroxide and nitric acid are used as cleaning agents. The spent cleaning fluids must be disposed of by suitable means. There can be emission of milk powder dust into the local environment during plant malfunctions but this is rare. Noise is a problem mainly within the plant buildings but fans can affect close neighbours. Unbleached paper and plastic laminate packaging must be disposed of in overseas markets.

Applications

It manufactures a wide range of spray dried milk powders (> 100) to meet the diverse and special needs of customers. Milk powders may vary in their gross composition (milkfat, protein, lactose), the heat treatment they receive during manufacture, powder particle size and packaging. Special "high heat" or "heat stable" milk powders are required for the manufacture of certain products such as recombined evaporated milk. Some powders are agglomerated and they may be instantised for easy use in the home. Instant powders must wet, sink and disperse quickly, with minimal stirring, when added to water. The resulting liquid should closely resemble fresh milk and be free from undissolved particles. Some powders are fortified with vitamins and minerals. It is very important to use a powder suited to the intended application. Milk powders of various types are used in a wide range of products including the following:

- Baked goods, snacks and soups.
- Cheese milk extension (powder is added to local fresh milk to increase the yield of cheese).
- Chocolates and confectionery (e.g. milk chocolate).
- Dairy desserts.
- Direct consumer use (home reconstitution).
- Ice cream.
- Infant formulae.
- Nutritional products for invalids, athletes, hospital use etc.
- Recombined "fresh", UHT, evaporated and sweetened condensed milks.
- Recombined cheeses, mainly "soft" or "fresh".
- Recombined coffee and whipping creams.
- Recombined yoghurts and other fermented products.

SPRAY DRYING OF DAIRY PRODUCTS

Milk is extremely perishable and yet, for a number of reasons, it has to be preserved for later consumption. The removal of water prevents the growth of micro-organisms and facilitates preservation and storage of milk constituents. Spray drying is one of the most convenient techniques for producing milk powders and for stabilising milk constituents.

Before spray drying, the milk undergoes several processes (such as heat treatment, cream separation, membrane processes, vacuum evaporation and homogenisation). The chemical, physical, technological, nutritional, functional and microbiological properties of the final products are influenced by a number of factors such as operating conditions, properties of the dairy products and storage conditions.

Dehydration of milk and whey is intended to stabilise these products for their storage and later use. These powders are used mostly in animal feeding. With changes in agricultural policies (such as the implementation of the quota system and dissolution of the price support system), the dairy industry was forced to look for better use of the dairy surplus and the by-products of cheese (whey) produced from milk and butter milk manufactured from cream. Studies were undertaken into reuse of protein fractions whose nutritional qualities and functioning led us to believe they could have multiple applications. The results have been a change in the nature of dairy powders during the last 15 years1. The total quantity of powder did not vary (approximately 1,400,000 t in 1986 and in 2004) but the amount of milk powder decreased by 40% between 1986 and 2004. This decrease mainly involved skimmed and full fat milk powders. However, the production of whole milk powder and whey powder increased by 62% and 73%, respectively, between 1986 and 2004. This increase was reflected in types of whey and derived powders (protein concentrates) produced. This overall change may be explained by the quota system implemented that gives better control over dairy production. As cheese production from cow's milk increased by 31% between 1986 and 2003, there was a corresponding increase in whey production and a decrease in milk powder production.

Since the 1980s, the dairy industry has developed new technological processes for extracting and purifying proteins (e.g. casein, caseinates, whey proteins etc.), such as dairy proteins and whey concentrates, micellar casein concentrates, native phosphocaseinate suspension (NPCS), selectively demineralised concentrates6 and super-clean skimmed milk concentrates – mainly due to advances in filtration technology (microfiltration, ultrafiltration, nanofiltration and reverse osmosis).

Most of these proteins are marketed in dehydrated form and used either as nutritional or functional ingredients. Milk 'cracking' in different dried and stable forms led to a sudden increase in the use of intermediate dairy products. Many new uses of these constituents then appeared with the manufacture of formula products, substitutes and adapted raw materials.

The most frequently used technique for dehydrating dairy products is spray drying. Different methods of biological stabilisation in which spray drying is a stabilisation method based on biological inhibition and reduction of water activity. Spray drying became popular in the industrial world in the 1970s, but at that time there were few scientific or technical studies about the practice and, in particular, none on the effects of spray drying parameters, nor the physico-chemical composition or microbiology of the concentrates on powder quality. Manufacturers acquired expertise in milk drying and eventually in whey drying processes through trial and error. Due to the variety and complexity of the mixes to be dried, a more rigorous method, based on physico-chemical and thermodynamic properties, has become necessary. Greater understanding of the biochemical properties of milk products before drying, water transfer during spray drying, properties of powders and influencing factors has now become indispensable in the production of milk powder. The lack of technical and economic information and of scientific methods has prevented manufacturers from optimising equipment in terms of energy costs and powder quality.

Properties of Spray Dried Milk Products

A dairy powder is not only characterised by its composition (proteins, carbohydrates, fats, minerals and water) but also by its microbiological and physical properties (bulk and particle density, instant characteristics, flowability, floodability, hygroscopicity, degree of caking, whey protein nitrogen index, thermostability, insolubility index, dispersibility index, wettability index, sinkability index, free fat, occluded air, interstitial air and particle size) which form the basic elements of quality specification. There are well-defined test methods for determining these characteristics, according to international standards. These characteristics depend on technological operations before drying, drying parameters (type of tower spray dryer, nozzles/wheels, pressure, agglomeration, thermodynamic conditions of the air: temperature, relative humidity and velocity) and characteristics of the concentrate before spraying (composition/physico-chemical characteristics, viscosity, thermo-sensibility, availability of water). The nutritional quality of dairy powders depends on the intensity of the thermal processing during the technological process. Thermal processing induces physicochemical changes which tend to decrease the availability of the nutrients (loss of vitamins, reduction of available lysin content, whey protein denaturation) or to produce nutritional compounds such as lactulose.

Modelling, Simulation and Water Transfer

The physical and biochemical qualities of milk powder depend on the water distribution in the concentrate at the air/water droplet interface, which, in turn, depends on the composition of the concentrate. Roos and Schuck et al. have described the physico-chemical properties of pure and bound water and the effects of water on physical state, transition temperatures, sticking temperature, reaction kinetics and stability of milk products. The emphasis is on the physical state of non-fatty solids and the effects of water and its physical state on physico-chemical changes, growth of micro-organisms and stability. Spray-drying, storage and quality of milk powder are significantly dependent on both the physical state of the lactose (one of the main components in skimmed milk powder) and on the proteins and other carbohydrates, which themselves are dependent on the glass transition temperature (Tg). The spray-drying of skimmed milk concentrate is so rapid that the lactose cannot crystallise. Rapid removal of water in subsequent spray drying does not allow lactose crystallisation and, when water is removed, lactose is transformed to a solid-like, amorphous glass directly from the dissolved state. Many dehydrated foods contain amorphous components in the glass form. This is a non-equilibrium state with higher energy in relation to the corresponding equilibrium state. If the temperature of a material in this state rises above a certain critical value then it transforms into a rubber. This phenomenon is known as glass transition and the temperature at which it occurs is the glass transition range temperature. This transition results in an increase in mobility in the rubbery states which, in turn, can produce changes in physical and chemical properties of the material. Carbohydrates, including sugars, starch and hemi-celluloses, can exist in the amorphous glassy state in dried foods. Low molecular weight sugars in the glassy state are usually extremely hygroscopic and have low glass transition temperatures. This can lead to problems in spray drying and/or in storage. The addition of high molecular weight compounds to these low molecular weight sugars can improve spray drying and storage conditions. Proteins, including gelatin, elastin, gluten, glutenin, casein, whey proteins and lysozyme, are also found in the amorphous state in dried food. In the dry state, they have a relatively high glass transition temperature. However, the complexity of the mathematical models presented makes it difficult for manufacturers to put them into practice.

Dairy Powders

The different processes of spray drying, combined with membrane filtration, also affect the physico-chemical environment, the purity and the biochemical properties of milk powders and produce a range of powders with different physical and functional properties such as high milk protein powder, whey protein powder, whole milk powder or high fat powder. Physico-chemical factors currently help the dairy industry to optimise drying parameters and characterise these new dairy concentrates (except for the dry matter and viscosity). The methods generally used to analyse solubility, dispersibility and wettability of milk powders provide unsatisfactory results because they do not fully take into account the new functions.

Equipment and Energy Consumption

For more than 30 years, spray drying has been the most frequently used milk drying technique. It is also the most convenient technique for producing powders directly from pumpable feeds. Indeed, since the 1970s there has been an increase in the capacity of tower spray dryers (from 1 to 6 t of water drained per hour). Tower spray dryers treating from 10 to 15 t of water per hour have recently been installed in New Zealand and in Australia. The total capacity and number of tower spray dryers have more than doubled in a short time in certain countries.

Spray drying involves atomising the feed into a spray of droplets which are put into contact with hot air in a drying chamber. There are three modes of contact: co-current, counter-current and mixed flow. Sprays are produced by a rotary (wheel) or nozzle atomiser.

The tower is a one stage spray drying unit which means that the processing time in the spray drying chamber is very short (approximately 20 – 60 s). There is thus no real equilibrium between air humidity and product humidity. Therefore, if the outlet air temperature is raised, the energy efficiency of the unit decreases. The two stage spray dryer consists of limiting spray drying for longer processes and is therefore closer to thermodynamic balance. Upon being discharged from the spray drying unit, the product should have a maximum moisture that is compatible with continuous evacuation. This significantly lowers the outlet air temperature and increases the inlet air temperature. In order to obtain the required residual moisture, the final drying takes place in an external vibrating fluid, or vibro-fluidiser, in which the air flow and the treatment temperatures are lower than in the chamber and thus better designed for qualitative preservation of the powder. The two stage dryer has demonstrated how to reduce drying costs and improve the performance of units, by transferring most of the drying from atomisation to phase fluidisation until the wet product begins to stick to the walls of the chamber. This contact is inevitable in view of the internal agitation necessary for the thermal exchange. Removing this limit led to complete overhaul of the spray drying phase and resulted in the three-stage dryer the biggest breakthrough in this field since the emergence of spray drying. As it is impossible to operate without the walls of the unit, the aim was to minimise any contact between the walls and the wet product using the three stage dryer. The latter was stabilised and dried in an internal fluid bed inside the spray drying chamber.

There are other spray dryer designs such as the tall-form tower, flat-bottom chamber, restricted-height chamber, high or extra high-temperature chamber, box dryer, integrated belt chamber (Filtermat) and integrated filter dryer (IFD). The type of tower spray dryer depends on the specific properties of the product to be dried (high fat content, starches, maltodextrin, egg products, hygroscopic products, etc.) and the choice of the technology used depends on the thermal efficiency (calculated according to different methods), the qualities and properties of the product to be dried and the powders to be obtained. A thermohygrometric sensor is used for some examples of such measurements (temperature, absolute and relative humidity, dry air flow rate, water activity), for calculation of mass and absolute humidity to prevent sticking in the dry chamber and to optimise powder moisture and water activity in relation to the relative humidity of the outlet air.

Recommendations for Fire Prevention in Spray Drying of Milk

Fire in spray dryers for milk or milk products can lead to dangerous situations for the operators and may cause serious damage to plant and buildings. Fire prevention in such areas is primarily achieved through efforts to avoid situations involving a fire hazard. Any situation that may involve a fire hazard must therefore be quickly detected, either through a system of recording and automatic alarms, or through visual inspection. In the event that a fire breaks out in spite of all precautions, provisions to avoid injury to personnel must be in place, to prevent and limit damage.

STERILIZATION

Sterilization of milk is aimed at killing all microorganisms present, including bacterial spores, so that the packaged product can be stored for a long period at ambient temperature, without spoilage by microorganisms. Since molds and yeasts are readily killed, we are only concerned about bacteria. To that end, 30 min at 110 °C (in-bottle sterilization), 30 sec at 130 °C, or 1 s at 145 °C usually suffices. The latter two are examples of so-called UHT (ultra-high-temperature, short time) treatment. Heating for 30 min at 110 °C inactivates all milk enzymes, but not all bacterial lipases and proteinases are fully inactivated; it causes extensive Maillard reactions, leading to browning, formation of a sterilized milk flavor, and some loss of available lysine; it reduces the content of some vitamins; causes considerable changes in the proteins including casein; and decreases the pH of the milk by about 0.2 unit. Upon heating for 1 s at 145 °C chemical reactions hardly occur, most serum proteins remain unchanged, and only a weak cooked flavour develops. It does not inactivate all enzymes, e.g., plasmin is hardly affected and some bacterial lipases and proteinases not at all, and therefore such a short heat treatment is rarely applied.

The undesirable secondary effects of in-bottle sterilization like browning, sterilization flavor, and losses of vitamins can be diminished by UHT sterilization. During packaging of UHT-sterilized milk, contamination by bacteria has to be rigorously prevented. After UHT sterilization, certain enzymatic reactions and physicochemical changes still may occur.

Sterilized milk may be defined as (homogenized) milk which has been heated to a temperature of 100°C or above for such lengths of time that it remain fit for human consumption for at least 7 days at room temperatures. Commercially sterilized milk is rarely sterile in the strict bacteriological sense. This is because the requirement for complete sterility conflict with the consumer preference for normal color and flavor in the sterilized product. The spore –forming bacteria in raw milk, which are highly heat-resistant, survive the sterilization temperature-time employed in the dairy and ultimately lead to the deterioration of sterilized milk.

The term "sterilization" when used in association with milk, means heating milk in sealed container continuously to a temperature of either 115 °C for 15 minutes or at least 130 °C for a period of 1 second or more in a continuous flow and then packed under aseptic condition in hermetically sealed containers to ensure preservation at room temperature for a period not less than 15 days from the date of manufacture.

Sterilization of foods by the application of heat can either be in sealed containers or by continuous flow techniques.

Sterilized milk is kept for a long time so that it will show extensive gravity creaming if unhomogenized. Creaming as such is undesirable. Besides, partial coalescence of the closely packed fat globules will lead to formation of a cream plug, which is hard to mix throughout the remaining milk; oiling off may occur at somewhat elevated temperatures. Therefore, sterilized liquid milk is always homogenized.

Advantages

- Remarkable keeping quality; does not need refrigerated storage.
- No cream layer/plug.
- Forms a soft digestible curd, and hence useful for feeding of infants and invalids.
- Distinction rich flavor (due to homogenization).
- Economical to use.
- Less liable to develop oxidized taints.

Disadvantage

- Increased cost of production.
- More loss in nutritive value than pasteurization.
- Gerber test by normal procedure not so accurate.

Sterilized Milk Must

- Keep without deterioration, i.e., remain stable and be of good commercial value for a sufficient period to satisfy commercial requirement.

- Be free of any micro-organisms harmful to consumer health, i.e., pathogenic toxinogenic germs and toxins.

- Be free of any micro-organisms liable to proliferate, i.e. it should not show signs of bacterial growth (which leads, inter alia, to an absence of deterioration).

In-bottle Sterilization

The raw milk, on receipt, should be strictly examined by the physic-chemical and bacteriological test and only high quality milk should be used for production of sterilized milk. Care should be taken to accept milk supplies which have no developed acidity and which contain the least number of spore-forming bacteria. The intake milk should be promptly cooled to 5 °C for bulk storage in order to check any bacterial growth. Next, it should be pre heated to 35-40 °C for efficient filtration/ clarification, so as to remove visible dirt, etc., and to increase its aesthetic quality. The milk should again be cooled to 5 °C so as to preserve its quality. It should then be standardized to the prescribed percentage of fat and solids-not-fat content in order to conform to legal standards. It must be stored at 5 °C until processing. The milk should be promptly pre heated to 60 °C for efficient homogenization to prevent any subsequent formation of a cream layer; usually single-stage homogenization is carried out at 2500 psi pressure. The homogenized milk must be clarified so as to remove the sediment formed during the homogenization process. The hot milk from the homogenizer should be filled into the cleaned and sanitized bottle coming from the bottle washing machine and then sealed with special caps. The filled and capped bottles should then be placed in metal crates for sterilization by the batch process, or fed into conveyors for the continuous process. Usually the milk is sterilized at 108-111 °C for 25-35 minutes. The sterilized milk bottles should be gradually cooled to room temperature. Any sudden cooling may led to bottle breakage. Finally the milk-in-bottles should be stored in a cool place.

Batch Sterilizers

These may either be rotary or non-rotary in type. The batch sterilizers are rectangular, horizontal, boiler shaped retorts with a steam inlet and condensate outlet, fitted with clamp-down covers, into which steam is adjusted for the required temperature and time for sterilization.

Advantages

- Simplicity and flexibility of operation.

- Less initial capital and recurring expenditure.

Disadvantages

- Usually produces a brownish appearance and cooked taste in the finished product.

- Sterilization may be faulty.

- Cooling has to be slow to avoid breakage.

- Economic advantages of large-scale processing are not obtained.

In the batch-rotary type, the filled bottles are put in to holders which are rotated at 6-7 rpm. The sterilized milk is of a slightly better quality in rotary-type sterilizers than in non-rotary ones.

Continuous Sterilizers

In this type, the filled and sealed milk bottles are automatically placed by means of a slat conveyor in to the pockets of carrier cages. They then passed into water at or near boiling temperature; from there, they enter the sterilizing zone, which consists of a steam chamber at 108-111 °C. Here the bottles remain for a pre-determined time, viz., 25-30 minutes, for milk sterilization

Cooling

After heat treatment in the batch/tank sterilizers, the milk bottles may be cooled in air or water. If cooling is too rapid, the bottles may crack; if too slow, there is a danger of browning due to caramelization. In the continuous system, after leaving the sterilizing zone, the bottles enter a column of hot water where the cooling process begins. This is followed by their passage through another tank of water for further cooling, and lastly through a shallow tank of cold water for final cooling. The bottles are then automatically discharged and conveyed to a point where they are placed in crates in which they are transferred to the storage room.

Turbidity Test for Sterilized Milk

The turbidity test depends upon the denaturation of proteins of milk especially albumin after sterilization. When solution of inorganic salts or acids are added albumin separates with the casein. The sample after treatment with ammonia sulphate is filtered and heating of filtrate shows turbidity due to presence of albumin on account of sufficient heat treatment. If milk has been sterilized properly all albumin will have been precipitated and no turbidity will be produced. The test is not suitable for UHT milk.

Procedure

Pipette 20 ml of milk in a 50 ml conical flask, add 4.0±0.1 g of ammonium sulphate. Shake the flask till the ammonium sulphate is completely dissolved. Allow the mixture to settle for 5 minute, filter through a folded filter paper in a test tube. Keep about 5 ml of the above filtrate in a boiling water bath for 5 min. Cool the tube in a beaker of cold water and examine the contents for turbidity by moving the tube in front of an electric light shaded from the eyes of the observer.

Interpretation

The milk is considered sterilized when the filtrate shows no turbidity.

Ultra High Temperature (UHT) Processing

More recently, continuous sterilization processes have been introduced. UHT or aseptic processing involves the production of a commercially sterile product by pumping the product through a heat exchanger. To ensure a long shelf life the sterile product is packed into pre-sterilized containers in a sterile environment. An airtight seal is formed, which prevents re-infection, in order to provide a shelf life of at least three months at ambient temperature. It has also been known for a long time that the use of higher temperatures for shorter times will result in less chemical damage to important nutrients and functional ingredients within foods, thereby leading to an improvement in product quality.

In these processes, the milk is heated to 135-150 °C for a few seconds, generally in a plate or tubular heat-exchanger. The milk, which is then almost sterile, has to be filled into containers for distributions; the filling has to be done aseptically. Ideally, heating and cooling should be as quick as possible.

This applies only as long as the product remains under aseptic conditions, so it is necessary to prevent re-infection by packaging the product in previously sterilised packaging materials under aseptic conditions after heat treatment. Any intermediate storage between treatment and packaging must take place under aseptic conditions. This is why UHT processing is also called aseptic processing.

UHT Plants

UHT plants are often flexibly designed to enable processing of a wide range of products in the same plant. Both low-acid products (pH > 4.5) and high-acid products (pH < 4.5) can be treated in a UHT plant. However, only low-acid products require UHT treatment to make them commercially sterile. Spores cannot develop in high-acid products such as juice, and heat treatment is therefore intended only to kill yeast and moulds. Normal high temperature pasteurisation (90-95 °C for 15-30 seconds) is sufficient to make high-acid products commercially sterile.

Various UHT Systems

There are two main types of UHT systems on the market.

Direct UHT Plants

In the direct systems the product comes in direct contact with the heating medium, followed by flash cooling in a vacuum vessel and eventually further indirect cooling to packaging temperature.

The direct systems are divided into:

- Steam injection systems (steam injected into product).

- Steam infusion systems (product introduced into a steam-filled vessel).

UHT processing means commercial sterility to ensure food safety and long shelf life at ambient temperature. It entails heating the product to a specific temperature for a specific length of time. The higher the temperature, the shorter the time required to destroy micro-organisms. The more rapidly the product can be heated and then subsequently cooled down again, the less impact the process has on the chemical changes in the product, such as changes in taste, colour and even to some extent, nutritional value. The most effective way of achieving rapid heating is to mix high temperature steam directly with the product, followed by flash cooling in a vacuum vessel. This is called a direct system.

Flash cooling is an operation, which as well as cooling, also involves deaeration and deodorisation of the treated product. In addition, deaeration secures higher homogenisation efficiency and the deaeration will also positively influence the storage stability of the processed product in terms of preventing oxidation during storage.

The rapid heating and cooling explains why direct systems deliver superior product quality and are often chosen to manufacture heat sensitive products, such as premium quality market milk, enriched milk, cream, formulated dairy products, soy milk and soft ice mix, as well as dairy desserts and baby food.

Indirect UHT Plants

In many cases, products must not only be attractive and healthy to eat and drink, but also economical to manufacture, store and distribute. The most cost-effective method of UHT processing is indirect heating a heating method in which the processed product never comes into direct contact with the heating medium. There is always a wall in between. This technique applies to all types of heat exchangers.

In the indirect systems the heat is transferred from the heating media to the product through a partition (plate or tubular wall). The indirect systems can be based on:

- Plate heat exchangers,
- Tubular heat exchangers,
- Scraped surface heat exchangers.

Indirect UHT plants are a suitable choice for processing of milk, flavoured milk products, cream, dairy desserts, yogurt drinks and other non-dairy applications, such as juices, nectars and tea.

Indirect UHT plant based on plate heat exchangers.

This process solution is appropriate for products such as coffee cream and evaporated concentrated milk.

Indirect UHT Plant based on Tubular Heat Exchangers

A tubular system is chosen for UHT treatment of products with low or medium viscosity that may or may not contain particles or fibres. Soups, tomato products, fruit and vegetable products, certain puddings and desserts are examples of medium-viscosity products well suited to treatment in a tubular concept. Tubular systems are also frequently utilised when longer processing times are required for ordinary market milk products.

Indirect UHT Plant based on Scraped Surface Heat Exchangers

Scraped surface heat exchangers are the most suitable type for treatment of high-viscosity food products with or without particles.

Aseptic Packaging

Aseptic packaging has been defined as a procedure consisting of sterilisation of the packaging material or container, filling with a commercially sterile product in an aseptic environment, and producing containers that are tight enough to prevent recontamination, i.e. that are hermetically sealed. The term "aseptic" implies the absence or exclusion of any unwanted organisms from the product, package or other specific areas.

For products with a long non-refrigerated shelf life, the package must also give almost complete protection against light and atmospheric oxygen.

Loss of Nutrients during Sterilization

The nutritive value of pasteurized and UHT-sterilized milk changes little by the heat treatment and during storage. In-bottle sterilized milk shows a somewhat greater loss of nutritive value. Of special importance are the decrease of available lysine and the total or partial loss of some vitamins. Maillard reactions are responsible for the partial loss of lysine. They occur to some extent in UHT-sterilized milk during storage and in in-bottle sterilized milk during heating. The loss of lysine is not serious in itself because in milk protein, lysine is in excess.

The losses of vitamins mainly concern vitamin C and some five vitamins of the B group. Vitamins A and E are sensitive to light and/or oxidation, but mostly their concentrations do not decrease in sterilized milk. Losses of vitamins in milk should be evaluated relative to the contribution of beverage milk to the supply of these vitamins in the total diet. Especially the loss of vitamins B1, B2, and B6 are considered undesirable. The loss of vitamin C is generally of minor importance as such, but it may affect the nutritive value in other ways. The breakdown of vitamin C is connected with that of vitamin B12; moreover, vitamin C protects folic acid from oxidation.

Loss of vitamins during storage can largely be avoided if O_2 is excluded. Vitamins C and B9 may completely disappear within a few days if much O_2 is present. The loss is accelerated by exposure to light, with riboflavin (vitamin B2) being a catalyst. Most of the riboflavin disappears on long-term exposure to light.

HOMOGENIZATION

Homogenization refers to the process of forcing the milk through a homogenizer with the object of sub-dividing the fat globules. Homogenization has become a standard industrial process, universally practised as a means of stabilising the fat emulsion against gravity separation. Gaulin, who invented the process in 1899, described it in French as "fixer la composition des liquides" which means it makes liquid composition stable.

The purpose of homogenization is to disintegrate or finely distribute the fat globules in the milk, in order to reduce creaming. Homogenization primarily causes disruption of fat globules into much smaller ones. Consequently, it diminishes creaming and mayalso diminish the tendency of globules to clump or coalesce. Essentially, all homogenised milk is produced by mechanical means.

Homogenized Milk

Homogenized milk is milk which has been treated in such a manner as toensure breakup of the fat globules to such an extent that after 48 hours quiescent storage no visible cream separation occurs on the milk; and the fat percentage of the milk in the top 100 ml of milk in a quart bottle, or of proportionate volumes in containers of other sizes, does not differ by more than 10 per cent of itself from the fat percentage of the remaining milk as determined after thorough mixing.

Objectives of Homogenization

Homogenization results in milk or milk products in which the fat globules are reduced in size to-such an extent that no visible cream separation occurs in the milk. This process basically results in milk of uniform composition or consistency and palatability without removing or adding any constituents. Homogenization increases the whiteness of milk, because the greater number of fat globules scatters light more effectively. Homogenized milk is less susceptible to oxidized flavor, and the softer curd formed by it when entering the stomach aids digestion .

Homogenization is applied for any of the following reasons:

- Counteracting creaming: To achieve this, the size of the fat globules should be greatly reduced. A cream layer in the product may be a nuisance for the user, especially if the package is nontransparent.

- Improving stability toward partial coalescence: The increased stability of homogenized fat globules is caused by the reduced diameter and by the acquired surface layer of the fat globules. Moreover, partial coalescence especially occurs in a cream layer, and such a layer forms much more slowly in homogenized products.

- Creating desirable rheological properties: Formation of homogenization clusters can greatly increase the viscosity of a productsuch as cream. Homogenized and subsequently soured milk (e.g., yogurt) has a higher viscosity than unhomogenized milk. This is because the fat globules that are now partly covered with casein micelles in the aggregation of the casein micelles.

- Recombining milk products: At one stage of the process, butter oil must be emulsified in a liquid suchas reconstituted skim milk. A homogenizer, however, is not an emulsifying machine. Therefore, the mixture should first be pre-emulsified, for example, by vigorous stirring; the formed coarse emulsion is subsequently homogenized.

Homogenizer

Homogenizers are high pressure, reciprocating pumps each having a sanitary head upon which the homogenizing valves are mounted. Positive displacement pumps are necessary to supply the feed to the valve. Homogenizers generally have either three or five pistons, driven from a crank shaft through connecting rods.

This is a machine which causes the sub-division of fat globules. It consists of a high pressure through a narrow opening between the homogenizing valve and its seat; the fat globules in the milk are thereby sub-divided into smaller particles of more uniform size. The homogenizing valve is held down by a heavy pressure spring against the seat of the valve. The valve and its seat are made of extremely hard material (e.g. stelite) and the contact faces are carefully ground so that the valve sits accurately on its seat. Homogenizers are either single stage or double stage.

Operation of the Homogenizer

Homogenizers of the common type consist of a high-pressure pump that forces the liquid through a narrow opening, the so-called homogenizer valve.

The disintegration of the original fat globules is achieved by acombination of contributing factors such as turbulence and cavitation. This is accompanied by a four to six-fold increase in the fat/plasma interfacial surface area. The newly created fat globules are no longer completely covered with the original membrane material. Instead, they are surfaced with a mixture of proteins adsorbed from the plasma phase.

Homogenization is done by forcing all of the milk at high pressures through a narrow slit, which is only slightly larger than the diameter of the globules themselves. The velocity in the narrowest slit can be 100 to 250 m/s. This can cause high shearing stresses, cavitation and micro-turbulence. The globules become deformed, then become wavy and then break up.

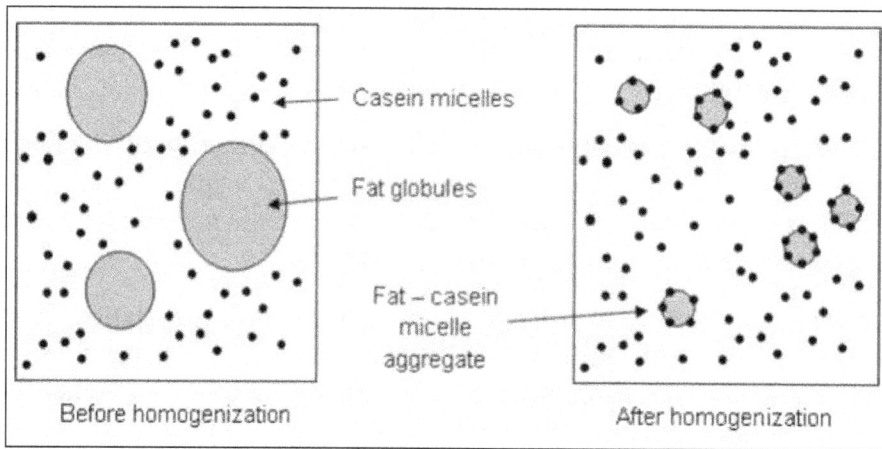

Effect of homogenization on fat and casein fractions in fluid milk.

Homogenization Theories

Many theories of the mechanism of high pressure homogenization have been presented over the years. For an oil-in-water dispersion like milk, where most of the droplets are less than 1 μm in diameter, two theories have survived. Together, they give a good explanation of the influence of different parameters onthe homogenising effect.

The theory of globule disruption by turbulent eddies ("micro whirls") is based on the fact that a lot of small eddies are created in a liquid travelling at a high velocity. Higher velocity gives smaller eddies. If an eddy hits an oil droplet of its own size, the droplet will break up. This theory predicts how the homogenising effect varies with the homogenising pressure. This relation has been shown in many investigations.

The cavitation theory, on the other hand, claims that the shock waves created when the steam bubbles implode disrupt the fat droplets. According to this theory, homogenization takes place when the liquid is leaving the gap, so the back pressure which is important to control the cavitation is important to homogenization. This has also been shown in practice. However, it is possible to homogenize without cavitation, but it is less efficient.

The Homogenizer in a Processing Line

In general, the homogenizer is placed upstream, i.e. before the final heating section in a heat exchanger. In production of UHT milk, the homogenizer is generally placed upstream in indirect systems but always downstream in direct systems, i.e. on the aseptic side after UHT treatment. In the latter case, the homogenizer is of aseptic design with special piston seals, sterile steam condenser and special aseptic dampers.

However, downstream location of the homogenizer is recommended for indirect UHT systems when milk products with a fat content higher than 6- 10 % and/or with increased protein content are going to be processed. The reason is that with increased fat and protein contents, fat clusters and/or agglomerates (protein) form at the very high heat treatment temperatures. These clusters/agglomerates are broken up by the aseptic homogenizer located downstream.

Full Stream Homogenization

Full stream or total homogenization is the most commonly used form of homogenization of UHT milk and milk intended for cultured milk products.

The fat content of the milk is standardized prior to homogenization, as is the solids-non-fat content in certain circumstances, e.g. in yoghurt production.

Partial Homogenization

Partial homogenization is used to save on energy and machinery. The milk is separated into skim milk and cream, and the cream is homogenized and mixed with the separated milk. Partial stream homogenization means that the main body of skim milk is not homogenized, but only the cream together with a small proportion of skim milk. This form of homogenization is mainly applied to pasteurised market milk. The basic reason is to reduce operating costs. Total power consumption is cut by some 80% because of the smaller volume passing through the homogenizer.

Single-stage and Two-stage Homogenization

Homogenizers may be equipped with one homogenising device or two connected in series, hence the names single-stage homogenization and two stage homogenization.

In both single-stage homogenization and two-stage homogenization, the whole homogenization pressure (P1) is used over the first device. In singlestage homogenization, the back pressure (P2) is created by the process. In two-stage homogenization the back pressure (P2) is created by the second stage. In this case the back pressure can be chosen to achieve optimal homogenization efficiency. Using modern devices, the best results areobtained when the relation P2/P1 is about 0.2. The second stage also reduces noise and vibrations in the outlet pipe.

Single-stage homogenization may be used for homogenization of products with high fat content demanding a high viscosity (certain cluster formation).

Two-stage homogenization is used primarily to reach optimal homogenization results and to break up fat clusters in products with a high fat content.

Factors Affecting Homogenization

- Temperature of homogenization: The milk should, at the time of homogenization, be at a temperature above the melting point of fat, viz., above 33 °C. This is because fat should be in the liquid state for proper subdivision. The enzyme lipase should be inactivated, preferably prior to homogenization or immediately afterwards. This can be achieved by heating the milk to a temperature of 55 °C. In routine practice, the milk is heated to 65-70 °C for homogenization. The danger zone for lipase activity, viz., temperature 38-49 °C, should be avoided during or after homogenization.

- Pressure of homogenization: In a single stage, up to 6 per cent fat milk, usually 2000-2500 psi pressure is sufficient. Higher pressures may increase the tendency for the milk to curdle when cooked, due to the increased destabilizing effect on milk-proteins. For liquid products with more than 6 per cent fat, two-stage homogenization is needed to prevent fat clumping: 2000 psi at the first stage and 500 psi at the second stage.

Effect of Homogenization

The effect of homogenization on the physical structure of milk has many advantages:

- Smaller fat globules leading to less cream-line formation.
- Whiter and more appetizing colour.
- Reduced sensitivity to fat oxidation.
- More full-bodied flavour, and better mouth feel.
- Better stability of cultured milk products.
- No formation of cream layer/plug.
- Produces soft curd and is better digested; hence recommended for infant feeding.

However, homogenization also has certain disadvantages:

- Increased cost of production.
- Increased sensitivity to light influences, leading to taste defects such as "rancid", "soapy" or "oxidized".
- The milk might be less suitable for production of semi-hard or hard cheeses because the coagulum will be too soft and difficult to dewater.
- Raw milk is unsuitable for homogenization as the flavour deteriorates rapidly due to lipase action.

Homogenization Efficiency

Homogenization must always be sufficiently efficient to prevent creaming. The result can be checked by determining the homogenization index, which can be found in the manner described in the following example:

A sample of milk is stored in a graduated measuring glass for 48 hours at a temperature of 4-6 °C. The top layer (1/10 of the volume) is siphoned off, the remaining volume (9/10) is thoroughly mixed, and the fat content ofeach fraction is then determined. The difference in fat content between thetop and bottom layers, expressed as a percentage of the top layer, isreferred to as the homogenization index. The index for homogenised milk should be in the range of 1 to 10.

Determination of Creaming Index

Low creaming index is an indication of good homogenization. Sterilized milk may be graded as under for the quality of homogenization, as described in BIS:

Quality of homogenization	Creaming index
Excellent	Upto 10
Good	11 to 20
Fair	21to 30
Bad	Over 30

Procedure

50 ml of milk sample at 20±10C added in three glass tubes (with outside diameter 24 mm, length with stopper 245 mm and graduated from 0 to 50 ml). Centrifuge for 15 min at 1000 rev/min. Using separate pipette, take 5 ml sample from upper part of tubes, carefully taking the cream that adheres to walls o the tube and transfer into a container (sample I). The empty the three tubes into a separate container (sample II). Measure the fat content of sample I and II by Gerber method.

Calculation:

$$Creaming\ index = \frac{A-B}{B} \times 100$$

Where,

A= Fat content of sample I, and

B= Fat content of sample II.

Farrall Index

The homogenization efficiency of milk also analysed using the Farrall index method as outlined by Trout. It is a microscopic method. A Farrall index ranging from 5 to 7 implied "excellent" homogenization efficiency. Index exceeding 10 indicates inefficient homogenization.

FLUID MILK PRODUCTION

Fluid milk is an industry term for milk processed for beverage use.

Milk, the lacteal secretion, practically free from colostrum, obtained from the complete milking of one or more healthy cows. Milk that is in its final package form for beverage use shall have been pasteurized or ultrapasteurized, and shall contain not less than 8.25% solids and not less than 3.25% milk fat. Milk may have been adjusted by separating part of the milkfat therefrom, or by adding thereto cream, dry whole milk, skim milk, or nonfat dry milk. Milk may be homogenized.

Milk solids are the non-water components of milk – protein , lactose, and minerals. Sometimes the combination of protein, lactose and minerals is called the solids not fat content, and when the fat is included it is called total solids content.

Standardization

The fat content of milk varies with species (cow, sheep, goat, water buffalo), animal breed, feed, stage of lactation, and other factors. In order to provide the consumer with a consistent product, most milk is standardized.

To achieve standardization, milk is processed through centrifugal separators to create a skim portion and a cream portion of the milk. Separation produces a skim portion that is less than 0.01% fat and a cream portion that is usually 40% fat, although the desired fat content of the cream portion can be controlled by changing settings on the separator. The cream portion is then added back to the skim portion to yield the desired fat content for the product. Common products are whole milk (3.25% fat), 2% and 1% fat milk, and skim milk (< 0.1% fat).

Pasteurization

The majority of fluid milk is pasteurized using a high temperature short time (HTST) continuous process of at least 161 °F (71.6 °C) for 15 seconds. These conditions provide fresh tasting milk that meets the requirements for consumer safety. Higher heat processes, such as ultrapasteurization or aseptic processing , are used to extend the shelf life of refrigerated products or allow for storage at room temperature, respectively, but may impart a cooked flavor to the milk.

Homogenization

The fat in milk is secreted by the cow in globules of non-uniform size, ranging from 0.20 to 2.0 μm. The non-uniform size of the globules causes them to float, or cream, to the top of the container. Milk that is not homogenized is sometimes referred to as "creamline" milk. Pasteurized milk does not necessarily need to be homogenized. However, homogenized milk should be pasteurized to inactivate native enzymes that deteriorate fat (lipases) and cause rancidity, which results in off-flavors and reduced shelf life in milk.

The purpose of homogenization is to reduce the milk fat globules size to less than 1.0 μm which allows them to stay evenly distributed in milk. Homogenization is a high pressure process that forces milk at a high velocity through a small orifice to break up the globules. The result of homogenization is the creation of many more fat globules of a smaller size. The native milk fat globules are covered in a protein membrane that stabilizes the fat phase in the aqueous (water) phase of milk. Although the milk fat globule membrane is disrupted during the homogenization process, it spontaneously migrates back to the fat globules after homogenization. The new globules created during homogenization are spontaneously coated by proteins in the skim phase from the original milk fat globules.

Vitamin Fortification

Fluid milk is often fortified with vitamin A and vitamin D. The package label must declare when milk is fortified.

Whole milk is considered a good source of vitamin A. Vitamin A is a fat soluble vitamin that is found in the fat phase of milk. The vitamin A content that occurs naturally in 2%, 1% and skim milk is less than in whole milk because of the lower fat levels. Nutritional concerns about consumption of lower fat milk in the late 1970s led to the required fortification of vitamin A in lower fat milks. To achieve the nutritional equivalence of whole milk, lower fat milks should be fortified to 300 IU vitamin A per 8 oz serving. The FDA encourages fortification to a level of 500 IU of vitamin A per 8 oz serving, which is 10 % of the recommended daily allowance (RDA).

Vitamin D is a fat soluble vitamin that occurs naturally in milk but at low levels. Because milk is not considered an important natural source of vitamin D in the diet, vitamin D fortification is voluntary. Fortification of milk with vitamin D began in the 1930s to reduce the incidence of rickets in children. Although rickets is not currently a major concern, adequate vitamin D is necessary for human health. Vitamin D helps with calcium absorption, has an important role in bone health, and has a protective effect in cancer. Milk may be fortified with vitamin D to a level of 100 IU per 8 oz serving, which is 25% of the RDA.

Specialty Milk Beverages

The dairy industry has developed specialty fluid milk beverages to meet the diverse nutritional needs of consumers. Lactose-reduced and lactose-free milk, and acidophilus milk were developed for people with lactose intolerance (maldigestion). Lactose-reduced and lactose-free milk are processed, prior to packaging, with the lactase enzyme to separate lactose into its component sugars, glucose and galactose. Acidophilus milk contains Lactobacillus acidophilus, a probiotic lactic acid bacterium that is beneficial to human health. The Lactobacillus acidophilus bacteria use lactose for an energy source and reduce the amount of lactose present in milk. They also make the lactase enzyme which assists humans with lactose digestion in the small intestine.

Specialty milk beverages are available that are tailored to specific segments of the population. There are milk beverages with added plant sterols aimed at helping to improve cholesterol levels and others that are fortified with protein and calcium designed for adults. There are carbohydrate-reduced and vitamin fortified milk beverages for people watching their weight. Milk beverages targeted for teen athletes are protein fortified and fat-reduced. Milk beverages designed for children are calcium fortified, fat-reduced and flavored. The flavored milks compete with soft drinks for children's attention and come in a wide range of flavors from the traditional chocolate and strawberry to milks flavored like their favorite candy bar or ice cream.

PASTEURIZATION

Pasteurization (or pasteurisation) is the process by which heat is applied to food and beverages to kill pathogens and extend shelf life.

The Purpose of Pasteurization

- To increase milk safety for the consumer by destroying disease causing microorganisms (pathogens) that may be present in milk.

- To increase keeping the quality of milk products by destroying spoilage microorganisms and enzymes that contribute to the reduced quality and shelf life of milk.

Pasteurization Conditions

Minimum pasteurization requirements for milk products are based on regulations outlined in the Grade A Pasteurized Milk Ordinance (PMO). These conditions were determined to be the minimum processing conditions needed to kill Coxiella burnetii, the organism that causes Q fever in humans, which is the most heat resistant pathogen currently recognized in milk. Milk can be pasteurized using processing times and temperatures greater than the required minimums.

Pasteurization can be done as a batch or a continuous process. A vat pasteurizer consists of a temperature-controlled, closed vat. The milk is pumped into the vat, the milk is heated to the appropriate temperature and held at that temperature for the appropriate time and then cooled. The cooled milk is then pumped out of the vat to the rest of the processing line, for example to the bottling station or cheese vat. Batch pasteurization is still used in some smaller processing plants. The most common process used for fluid milk is the continuous process. The milk is pumped from the raw milk silo to a holding tank that feeds into the continous pasteurization system. The milk continuously flows from the tank through a series of thin plates that heat up the milk to the appropriate temperature. The milk flow system is set up to make sure that the milk stays at the pasteurization temperature for the appropriate time before it flows through the cooling area of the pasteurizer. The cooled milk then flows to the rest of the processing line, for example to the bottling station. There are several options for temperatures and times available for continuous processing of refrigerated fluid milk. Although processing conditions are defined for temperatures above 200°F, they are rarely used because they can impart an undesirable cooked flavor to milk.

The process of heating or boiling milk for health benefits has been recognized since the early 1800s and was used to reduce milkborne illness and mortality in infants in the late 1800s. As society industrialized around the turn of the 20th century, increased milk production and distribution led to outbreaks of milkborne diseases. Common milkborne illnesses during that time were typhoid fever, scarlet fever, septic sore throat, diptheria, and diarrheal diseases. These illnesses were virtually eliminated with the commercial implementation of pasteurization, in combination with improved management practices on dairy farms. In 1938, milk products were the source of 25% of all food and waterborne illnesses that were traced to sources, but now they account for far less than 1% of all food and waterborne illnesses.

Pasteurization is the process of heating a liquid to below the boiling point to destroy microorganisms. It was developed by Louis Pasteur in 1864 to improve the keeping qualities of wine. Commercial pasteurization of milk began in the late 1800s in Europe and in the early 1900s in the United States. Pasteurization became mandatory for all milk sold within the city of Chicago in 1908, and in 1947 Michigan became the first state to require that all milk for sale within the state be pasteurized. In 1924 the U.S. Public Health Service developed the Standard Milk Ordinance to assist states with voluntary pasteurization programs. The Grade A Pasteurized Milk Ordinance (PMO), as it is now called, is administered by the U.S. Departments of Health and Human Services and Public Health, and the Food and Drug Administration and defines practices relating to milk parlor and processing plant design, milking practices, milk handling, sanitation, and standards for the pasteurization of Grade A milk products. Each state still regulates milk processing within their own state but dairy products must meet the regulations stated in the PMO for products that will enter interstate commerce.

Initial pasteurization conditions, known as flash pasteurization, were to heat the milk to 155 to 178 °F (68.3 to 81 °C) for an instant followed by cooling. Pasteurization conditions were adjusted to 143 °F (61.7 °C) for 30 minutes or 160 °F (71.1 °C) for 15 seconds to inactivate Mycobacterium bovis, the organism responsible for tuberculosis. However, in 1957 these conditions were shown to be inadequate for the inactivation of Coxiella burnetii which causes Q fever in humans. New pasteurization conditions of 145 °F (62.8 °C) for 30 minutes for a batch process, or 161 °F (71.7 °C) for 15 sec for a continuous process, were adopted in order to inactivate Coxiella burnetii, and these conditions are still in use today.

OPTIMUM THERMAL PROCESSING FOR EXTENDED SHELF-LIFE (ESL) MILK

Extended shelf-life (ESL) or ultra-pasteurized milk is produced by thermal processing using conditions between those used for traditional high-temperature, short-time (HTST) pasteurization and those used for ultra-high-temperature (UHT) sterilization. It should have a refrigerated shelf-life of more than 30 days. To achieve this, the thermal processing has to be quite intense. The challenge is to produce a product that has high bacteriological quality and safety but also very good organoleptic characteristics. Hence the two major aims in producing ESL milk are to inactivate all vegetative bacteria and spores of psychrotrophic bacteria, and to cause minimal chemical change that can result in cooked flavor development. The first aim is focused on inactivation of spores of psychrotrophic bacteria, especially Bacillus cereus because some strains of this organism are pathogenic, some can grow at ≤ 7 °C and cause spoilage of milk, and the spores of some strains are very heat-resistant. The second aim is minimizing denaturation of β-lactoglobulin (β-Lg) as the extent of denaturation is strongly correlated with the production of volatile sulfur compounds that cause cooked flavor. It is proposed that the heating should have a bactericidal effect, B* (inactivation of thermophilic spores), of >0.3 and cause $\leq 50\%$ denaturation of β-Lg. This can be best achieved by heating at high temperature for a short holding time using direct heating, and aseptically packaging the product.

Extended shelf-life (ESL) milk has gained substantial market share in many countries. It has a refrigerated shelf-life of 21–45 days with some manufacturers claiming a shelf-life of up to 90 days. It is produced by two principal technologies: (1) Thermal processing using more severe conditions than pasteurization but less severe than ultra-high-temperature (UHT) processing; and (2) Non-thermal processes such as microfiltration and bactofugation, usually combined with a final thermal pasteurization treatment to meet regulatory requirements.

Heating Methods

The heating systems used for ESL processing are of two major types, direct and indirect. In direct systems, heating occurs through direct contact between steam and the product and in indirect systems the heat is transferred to the product from steam or hot water through a stainless steel barrier in a heat exchanger.

Direct Heating

In direct heating processes, the milk is first heated indirectly in a plate or tubular heat exchanger to 70–80 °C and then heated to the required high temperature by direct contact with dry culinary steam. The milk is held at the required temperature for the required period of time while it passes through a holding tube. The heated milk then passes to a vacuum chamber, which removes the water from the condensed steam and cools the milk to approximately the same temperature to which it was pre-heated prior to the steam heating stage. The milk is then cooled indirectly to ~4 °C.

There are two modes of steam heating, steam injection and steam infusion. In milk processing, these are often described as steam-into-milk and milk-into-steam. They differ considerably in terms of equipment but ESL milk produced by the two methods is very similar, although some authors have reported advantages of steam infusion over steam injection.

The major distinguishing feature of direct heating methods relevant to ESL milk processing is the high rate of heating and cooling, on the order of 0.5 s for a temperature change of 50–60 °C.

Indirect Heating

Indirect heating involves the use of plate or tubular heat exchangers for all heating and cooling stages. Of note in this regard is that a considerable amount of heat (up to ~90%) can be recovered by using the heat in the hot milk, after the holding tube, to heat the incoming cold milk. The heat recovery is greater than for direct systems, where it is ~50%. This is obviously an economic advantage of indirect systems.

The rates of heating and cooling in the high-temperature sections of indirect systems are much slower than in direct systems. This has an important consequence for ESL processing because, for the same bactericidal effect, the indirect systems cause more chemical change than do direct systems. Thus more cooked flavor is produced in indirect systems than in direct systems for the same bacterial, including spore, inactivation.

Nominal Temperature–Time Combinations

For direct systems, the only heat input of relevance to spore inactivation occurs between the preheat temperature (70–80 °C) and the highest temperature reached, both in the heating and cooling stages, and, because heating and cooling are so rapid, this heat input is very close to what occurs in the holding tube only. However, for indirect systems the heat input of interest includes that to which the milk is subjected in the heating and cooling stages as well as the holding tube.

Heating conditions are usually described in terms of a temperature–time combination, e.g., 125 °C for 5 s. This condition applies to the holding tube only and hence, for ESL milk, closely reflects the heat input in a direct heating system but considerably underestimates the heat input of interest in an indirect heating system. This was well illustrated by Rysstad and Kolstad who showed that for a direct and an indirect process having equal bactericidal effects (the nominal temperature–time combinations were 135 °C for 0.5 s and 127 °C for 1 s, respectively), the direct system caused much less chemical change than the indirect system. The lower temperature in the indirect system is compensated for by the extra heat input from the upper heating and cooling sections to achieve the desired bactericidal effect.

The bactericidal effect is usually expressed in terms of either B* or F_0. B* refers to inactivation of thermophilic spores; a process with B* = 1 causes a 9-log reduction of the spores and is equivalent to holding the product at 135 °C for 10.1 s. F_0 refers to inactivation of Clostridium botulinum spores; a process with F_0 = 2.8 causes a 12-log reduction of the spores and is equivalent to holding the product at 121.1 °C for 2.8 min.

The key aim for optimizing the heating conditions for ESL processing is to maximize the sporicidal effect but minimize the chemical effect, which results in cooked flavor production. For this, direct heating systems are more appropriate than indirect systems.

Alternative Direct Heating Systems

The Pure-Lac System

This system (SPX FLOW, Charlotte, NC, USA), based on steam infusion, was developed to produce ESL and UHT milk with a taste similar to that of pasteurized milk but to effect a high level of spore inactivation. The system also includes specially developed packaging equipment that ensures post-processing contamination (PPC) is minimal. The temperature–time conditions for the holding tube are given as 130–145 °C for <1 s. This is a very broad temperature range and the effects of heating at 130 °C and 145 °C are very different, with the lower temperature being marginal for inactivating spores. It is unclear what temperature was used to produce the ESL milk on which the bacteriological, chemical, and sensory analyses were performed. It states that the temperature for the Pure-Lac process can be "as high as 140 °C"; however, a temperature–time profile given for the Pure-Lac process indicates a temperature of 135 °C.

The short holding time is a central feature of the system. Precise control of this is achieved by imposing sufficient backpressure on the holding tube to ensure single-phase turbulent flow.

The Innovative Steam Injection (ISI) Process

The development of Innovative Steam Injection (ISI) heating, which can heat milk at 150–180 °C for <0.1 s. The process achieved a very high bacterial kill and was capable of producing commercially sterile milk with minimal chemical change; it caused only 20–25% denaturation of β-Lg. However, it did not inactivate the native milk protease, plasmin, and in UHT milk stored at 20 °C, bitterness due to plasmin-induced proteolysis resulted in a short shelf-life. However, for milk stored at 7 °C, no proteolysis occurred during storage for 28 days because plasmin has very low activity at this low temperature. Therefore ISI technology is applicable to ESL milk without an additional pre-heating step to inactivate plasmin, which is necessary for UHT milk.

The Millisecond Technologies (MST) Process

A patented technology involving both temperature and pressure manipulation has been proposed for extending the shelf-life of pasteurized milk. According to Millisecond Technologies Corporation, the commercial suppliers of the equipment, the process for extending the shelf-life of milk operates at <70 °C for <1 s and achieves a bacterial kill of ≥8 logs and a shelf-life of 45–60+ days. This was compared with HTST pasteurization at 72–80 °C for 15 s, which produces milk with a shelf-life of 14–18 days. However, Myer et al. reported somewhat different information. In their most successful trial, milk, pasteurized at 73.8 °C, entered an MST chamber at 73 °C and exited the chamber at 75.2 or 78.5 °C. Bacterial counts for these milks were reported to be <10 CFU/mL after storage for 63 days at 4 °C. However, Bacillus and Paenibacillus species were isolated after this treatment indicating that spores were not inactivated by the treatment.

According to the patent, the destruction of bacteria by the MST process is partly by heat and partly by the rapid "pressure variation" of at least 105 Pa/s. Milk is sprayed under pressure (800,000 Pa according to Myer et al.) in droplets of ≤0.3 mm into a heated MST chamber under vacuum (~0.025 Pa). The milk is rapidly heated in the chamber to the desired temperature (in <0.02 s). The bacterial kill is largely attributable to sudden decompression in the MST chamber. Rapid decompression has previously been reported to inactivate vegetative bacteria and even heat-tolerant spores.

MST technology has potential for processing milk although, as Myer et al. commented, several parameters of the process need to be examined including the efficacy of the process to inactivate more thermally robust microorganisms such as Bacillus species. Until the technology can be shown to inactivate spores of psychrotrophic spore-forming bacteria, it has limited application for production of ESL milk with a long shelf-life.

Microbiological Considerations

There are two major aspects to be considered when optimizing the conditions for producing ESL milk, microbiological and sensory. Ideally, ESL milk should not show any microbial growth during refrigerated storage. However, this is generally not the case because of several factors.

The microbiological issues can be divided into those related to the bactericidal nature of the heating conditions and those associated with post-processing contamination. The relative importance of these for the shelf-life of ESL milk depends on whether the milk is packaged aseptically or under very clean, but not aseptic, conditions. The nature of the heating conditions and storage temperature are the only factors relevant to the microbiology of aseptically packaged milk, while post-processing contamination together with the heating and storage conditions is important for ESL milks not packaged aseptically. Since ESL milk is usually not packaged under aseptic conditions, the bacteria in such ESL milk include spore-forming bacteria whose spores are not destroyed by the heating process, as well as spore-forming and non-spore-forming bacteria entering the milk after processing.

A further consideration that can affect ESL milk is the bacteriological quality of the raw milk. Because of the logarithmic reduction of bacterial counts by heat, the higher the bacterial count in the raw milk, the higher will be the residual count in the heated milk. Also, if psychrotrophic bacteria such as Pseudomonads are allowed to grow in the raw milk and produce heat-resistant proteases, bitter flavors can develop in the ESL milk during storage. As a rule of thumb, the total count in the raw milk should not exceed 105 CFU/mL.

Microbiological Issues Related to the Heating Process

Bacterial growth in ESL milk is by psychrotrophic organisms only as ESL milk is stored under refrigeration. Since ESL heating at ≥120 °C destroys all non-spore-forming bacteria but does not destroy all spores, psychrotrophic spore-formers are a major concern. They are the only bacteria that cause spoilage when post-processing contamination is eliminated, as in aseptically packaged milk, provided the storage temperature is kept at less than ~6 °C. The ESL heat treatment can also activate spores to germinate and thus allow their vegetative cells to grow during storage.

There have been a small number of investigations on the spore population of ESL milks collected or packaged under aseptic conditions. Aseptically collected milk samples from a commercial direct processing plant operated at 127 °C for 5 s. They found that B. licheniformis (the dominant organism at 73% of isolates), followed by B. subtilis, B. cereus, Brevibacillus brevis, and B. pumilus, were the most common spore-formers. When the ESL milks were held at 10 °C for 23 weeks, no growth occurred, indicating the absence of psychrotrophic spore-formers. However, when the milks were incubated at 30 °C, growth occurred. Bacteria isolated from the 30 °C incubated milks were found to be able to grow at 8 or 10 °C, suggesting some spores were sub-lethally injured during the heat treatment and were revived at 30 °C. B. cereus was the most psychrotrophic of the spore-formers isolated. Blake et al. found B. licheniformis, B. insolitus, B. coagulans, and B. cereus/thuringiensis in poor-quality milk (total count of >108 CFU/mL) directly heated at 120–132 °C for 4 s and packaged aseptically. No organisms grew in milk processed at temperatures ≥134 °C for 4 s. In a trial with good-quality milk, Blake et al. observed no psychrotrophic growth in milks processed at ≥128 °C. Although not dealing with ESL milk as it is known today, the study of Cromie et al. is instructive. They found that B. circulans was the dominant spoilage organism in milks heated at 72 to 88 °C for 15 s, aseptically packaged and stored at 3 or 7 °C for up to seven weeks. A similar observation was reported by Rysstad and Kolstad, where the total bacterial count of aseptically packaged pasteurized milk stored at 6 °C reached 106 CFU/mL (a nominal spoilage level) after 40 days. They also reported that when ESL milk produced by the Pure-Lac™ system was packaged aseptically, a shelf-life at the "relatively abusive temperature" of 10 °C of ≥45 days was achieved.

Psychrotrophic Spore-formers

In determining the ideal temperature time combinations of ESL milk processing to ensure a long shelf-life, a major consideration is choosing combinations that destroy spores of psychrotrophic bacteria. Spores of psychrotrophic bacteria are generally more heat-sensitive than those of mesophilic and thermophilic bacteria and hence less severe conditions will be required to destroy the former, although there are exceptions, e.g., some Paenibacillus species and some strains of B. cereus. In this regard, processing of ESL milk differs from UHT processing, in which the aim is to destroy the spores of mesophilic and thermophilic bacteria. In fact, the accepted minimum requirement for UHT processing to produce "commercially sterile" milk is a 9-log reduction of the thermophilic spore count; this is equivalent to heating at 135 °C for 10.1 s. For ESL milk, the suggested corresponding criterion is a 6-log reduction of the psychrotrophic spore count. However, the conditions used for producing ESL milk seldom meet this criterion.

The numbers and types of spores differ considerably between raw milk samples. This is largely due to the different environmental conditions of the cows as most spores enter the milk from teat and udder surfaces during milking. High spore numbers are found in very dry dusty conditions, wet and muddy conditions, and in the environs of housed animals, particularly when cows are fed silage. Thus the spore counts in raw milk from unhoused, pasture-grazing cows are typically <102 CFU/mL but 103 CFU/mL in milk from housed cows. Furthermore, the proportion of the spores that will germinate and grow under refrigeration conditions that is, the proportion that is psychrotrophic also varies. Reported percentages of raw milk samples containing psychrotrophic spore-formers vary from 25% to 83%.

Variation in the types of spores may be due to differences in environmental conditions. For example, carried out research on ESL milk in Logan, UT, USA, where the ambient temperatures are low, and suggested that their results on the heating conditions for inactivating psychrotrophic spores may have been influenced by acclimatization of the spores to cold environmental conditions. This may suggest that if conditions for producing ESL milk are set to those found by Blake et al. to produce ESL milk with a long shelf-life, they should be suitable for all ESL milk production and may in fact incorporate a margin of safety.

Paenibacillus and B. cereus are spore-formers of particular relevance to ESL milk.

Paenibacillus

Paenibacillus, which was formerly part of the genus Bacillus, has emerged in recent years as an organism of concern to the dairy industry. It has been found in silage and feed concentrates and can enter the milk from these sources. Doll et al. reported it to be the major organism in raw bulk tank milk (48% of isolates). It has also been reported to be a major spoilage bacterium in pasteurized milk in New York State and in ESL milk in South Africa. It can cause bitterness due to its production of proteases. Scheldeman et al. reported Paenibacillus spores in UHT milk, suggesting its resistance to UHT processing.

Paenibacillus species have a broad range of growth temperatures from ~5 to 55 °C. Their growth in pasteurized and ESL milk is evidence of their growth at low temperature, but Scheldeman et al. reported they could grow at temperatures as high as 55 °C. The optimum growth temperatures of Paenibacillus species range from 28 to 42 °C. Therefore, Paenibacillus can be psychrotrophic, mesophilic, or thermophilic and produce heat-resistant spores. This is an unusual combination of properties and an unfortunate one for the dairy industry as it indicates some strains of these organisms can survive all common milk heat treatments and grow at both refrigeration and ambient temperatures.

Cereus

Apart from the fact that the growth of psychrotrophic spore-formers can limit the shelf-life of ESL milk, some psychrotrophic spore-formers are pathogenic and hence pose a safety issue in ESL milk. The major organism of concern is Bacillus cereus, although some authors have also included other Bacillus species such as B. circulans as potential pathogens. B. cereus is a major spoilage organism in pasteurized and ESL milk. Some strains are pathogenic and the spores of some are quite heat-resistant.

There is a vast amount of information on this organism in the literature. The importance of B. cereus in the dairy/food industry is indicated by the reviews that have appeared on it, e.g.

- Incidence: B. cereus is widespread in the environment and is a common contaminant of milk and milk products. It has some closely related species; in fact, the B. cereus group (sometimes referred to as B. cereus sensu lato) comprises eight species: B. cereus (B. cereus sensu strict), B. anthracis, B. thuringiensis, B. mycoides, B. pseudomycoides,

B. weihenstephanensis, B. toyonensis, and B. cytotoxicus, which are difficult to distinguish phenotypically. It can therefore be assumed that in most cases where "B. cereus" has been isolated from milk and milk-based products, it could include other members of the B. cereus group. Using 16S rRNA gene sequence data of dairy isolates, Ivy et al. identified B. cereus, B. weihenstephanensis, and B. mycoides from the B. cereus group. It is interesting to note that B. thuringiensis has been isolated from UHT milk. This organism is used as a biological control agent and could be in high concentrations in some agricultural environments and animal feed sources. Furthermore, B. toyonensis is used as a feed additive.

Several surveys of the incidence of B. cereus in raw and pasteurized milk and milk-based products have been conducted, with the percentage of B. cereus-positive samples ranging from very low to 100%. Its incidence in raw and pasteurized milk is commonly 20–60%. Griffiths and Phillips reported that three studies found it to be the main psychrotrophic spore-former in raw milk and others have found it to be the dominant psychrotrophic spore-former in pasteurized milk, e.g., In general, the counts of B. cereus in raw milk are low, <100 CFU/mL, often <1 CFU/mL.

Like other spore-formers, B. cereus can enter milk from numerous sources. On farms, the main sources are water, cows' udders and teat surfaces, dust, soil and milkstone deposits on farm bulk tanks and pumps, pipelines, and gaskets and processing equipment in factories. Stewart stated that the cleaning and sterilizing systems used in equipment are not very effective at eliminating spores and may even activate spores of B. cereus. In the factory, fouling deposits and stainless steel surfaces to which the spores can readily attach are the main sources. Franklin reported a case of a very heat-resistant B. cereus spore that contaminated UHT cream (processed at 140 °C for 2 s) being traced to an upstream homogenizer. At the factory, B. cereus may enter ESL milk from processing lines containing dead ends, pockets, corners, crevices, cracks, and joints due to the ability of their spores to readily attach to stainless steel, glass, and rubber. When attached to stainless steel, spores show enhanced resistance to cleaning solutions. Mugadza and Buys reported that Paenibacillus, B. pumilus, and B. cereus were isolated from both filler nozzles and ESL milk; in the case of B. cereus there was a close relationship between isolates from the ESL milk and those from the filler nozzles. This indicates the importance of biofilms on equipment in the contamination of milk by B. cereus. Biofilms are surface-associated multicellular microbial communities embedded in an extracellular polysaccharide matrix; they are difficult to remove by normal cleaning procedures.

Several authors have reported a seasonal effect on the incidence of B. cereus spores in milk. However, the reports are inconsistent. Where cows are housed during the winter, high spore counts are often encountered and attributed to contamination from bedding and fodder. However, observed higher counts during spring and summer when cows are not housed. Stewart proposed that this may be due to the greater amount of dust during the summer. Slaghuis et al. reported that the milk from cows grazing on pasture during summer contained more B. cereus than milk from cows that were housed and fed conserved feed; this suggests that pasture may be a reservoir for B. cereus spores.

- Pathogenicity: The major interest in B. cereus from a public health perspective is related to the production of enterotoxins by several strains and its potential to cause two types of illness, diarrheal and emetic syndromes. The two types of toxins are very different. The diarrheagenic toxins are proteins with molecular weights in the range 38,000–46,000 Da.

They are produced by actively growing cells and are thermolabile, being inactivated by heating at 56 °C for 30 min. Two of three forms of the diarrheagenic toxins are believed to cause food poisoning in humans. In contrast, the emetic toxin is a cyclic peptide, cereulide, with a molecular weight of 1200, which is extremely resistant to heat, surviving heating at 126 °C for 90 min. Psychrotrophic strains of B. cereus growing at low temperature do not produce the emetic toxin but may produce the diarrhoeagenic toxins, albeit slowly and in low concentrations.

Despite its widespread presence in milk and milk-based products, B. cereus has been implicated in very few cases of illness; however, one outbreak in the Netherlands in which pasteurized milk was implicated involved 280 patients. Several authors have sounded a warning of the potential for this organism to cause disease although it has been suggested that toxin-producing strains of B. cereus in milk and milk products are unlikely to cause food poisoning as their production of toxin, even at high counts, is very low. Another reason why B. cereus rarely causes food poisoning is because it produces an intensely bitter flavor, making the contaminated products organoleptically unacceptable before they become toxic.

Te Giffel et al. found that 28 of 37 isolates from pasteurized milk produced enterotoxin. Strains that fermented lactose produced more enterotoxin than strains that did not. Van Netten et al. found that 25% of psychrotrophic B. cereus isolates from pasteurized milk were enterotoxin-positive. Notermans et al. stated that ≥ 105 CFU/mL of toxigenic B. cereus in pasteurized milk is generally considered to be a health hazard. They estimated that such numbers could be present in 7% of milk in the Netherlands at the time of consumption. However, epidemiological evidence does not indicate that B. cereus in milk causes disease to anywhere near this extent and hence the dose–response relationship needs to be revisited. Since the growth rate and enterotoxin production of B. cereus is low at 4 °C, several authors have concluded that milk or milk-based products stored at or below this temperature present a very low risk of becoming toxic, provided products are not stored for unduly long periods of time, e.g., >20 days. Since ESL milk is designed to have a shelf-life of at least 30 days, it is possible for B. cereus to reach high counts by the end of its shelf-life if it is not destroyed by the heat process or the ESL milk is not packaged aseptically and is contaminated with the organism after the heat treatment.

- Growth temperatures: B. cereus can grow at a range of temperatures but the optimum growth temperature is generally 30–37 °C. The maximum temperature for most strains is 45–50 °C. However, some strains are capable of growing at low temperatures and these are of most concern for ESL milks. They are termed psychrotrophic (able to grow at 7 °C) or psychrotolerant (able to grow at 4 °C but not at 43 °C). The species name, B. weihenstephanensis, has been used for this sub-group of B. cereus.

In surveys, the percentage of strains capable of growing at 7 °C has varied. For example, isolated 766 B. cereus strains from farm environments and raw milk and found the percentage of isolates capable of growing at 7 °C was 40% and 30%, respectively. Similarly, in a survey of pasteurized milk samples, found that 53% of 106 isolates tested were psychrotrophic. However, in a survey of milk from a fluid milk processing plant and a milk powder plant, these authors found only 6% of isolates from the first and no isolates from the second plant were psychrotrophic. An interesting phenomenon observed by Mayr et al. was that B. cereus (and three other spore-formers) grew at 8 °C after culturing at 30 °C but had not previously grown in milk at 10 °C.

For strains capable of growing at <7 °C, their growth rates decrease considerably as the temperature is decreased. Rowan and Anderson reported that of 38 psychrotrophic B. cereus isolates from milk-based infant formulae, one, four, and 16 isolates showed growth after 15 days at 4, 6, and 8 °C, respectively. Dufeu and Leesement reported the average generation times for the four strains to be 1.3 h at 30 °C, 9.1 h at 8 °C, and 54 h at 3 °C, respectively. This compares favorably with 9.4 to 75 h (average 8.2 h) at 7 °C.

- Spoilage potential: As well as being potentially pathogenic, B. cereus can cause substantial spoilage. It produces protease, which causes sweet curdling and bitterness in pasteurized milk. The defects were noticed after 8–10 days of storage at 5–7 °C. B. cereus also produces phospholipase C (sometimes referred to as lecithinase), which degrades phospholipids of the milk fat globule membrane and causes fat globule coalescence, or chemical churning, resulting in defects such as bitty cream. Lewis commented that because the growth of B. cereus is accompanied by the production of a disagreeable odor and flavor, consumers are likely to detect spoilage well before the milk becomes a safety issue.

- Heat resistance: The heat resistance of B. cereus is particularly relevant to ESL milk; however, much of the information in this section is also applicable to psychrotrophic spore-formers in general. While the spores of many psychrotrophic strains are not very heat-resistant, there is actually a wide range of heat resistance amongst the spores of B. cereus isolates, with some strains being highly heat-resistant. Mikolajcik reported that B. cereus (and B. licheniformis) produced the most heat-resistant spores in milk. Franklin and Vyletelova et al. showed that some B. cereus spores survive UHT treatment.

Heat resistance data, D- and z-values, have been reported for the spores of several B. cereus spores. Unfortunately, different researchers have determined the D-values at different temperatures, which makes comparison difficult. The range of reported heat resistance data is illustrated by the following D-values in 21 studies that were collated by Bergere and Cerf: D90, 3.6–10.8 min; D95, 0.5–20.2 min; D100, 0.3–27 min; D105, 11.2 min; D110, 11.5 min; D121, 0.03–0.04 min;. Stoeckel et al. later collated D-value data from five subsequent reports, including their own on infant formula; the range of D-values were as follows: D90, 1.1–12.8 min; D95, 2.0–4.4 min; D100, 0.27–1.83 min; and D110, 0.05–0.6 min. These values do not include the D values they obtained for spores in concentrated (50% total solids) infant formula, which were about double those of spores in standard (10% total solids) reconstituted infant formula. Van Asselt and Zweitering collated 465 data points from 12 publications and determined the mean D120-value of B. cereus spores to be 0.041 min (2.46 s) and the upper 95% prediction interval D120-value to be 0.52 min (31.2 s).

In about half of the studies reviewed by Bergere and Cerf, the heat inactivation curves were not linear; two showed shoulders and 10 showed tails. The importance of this is exemplified in a very heat-resistant B. cereus isolate from UHT cream that originated from an upstream homogenizer; although the majority of the spores were destroyed at 95–100 °C, a resistant fraction of ~1 in 105–106 survived heating at 135 °C for 4 h. This small resistant fraction was sufficient to cause contamination of the UHT cream. Bradshaw et al. also reported a very heat-resistant strain of B. cereus, but in this case no heat-resistant tail was observed. Its spores had D-values as follows: D115.6, 11.4 min; D121.1, 2.3 min; D126.7, 0.3 min and D129.4, 0.24 min. A report by Dufrenne et al. gave the range of D90-values for spores of 11 B. cereus isolates as 2.2–9.2 min but one other strain had a D90-value of >100 min. Therefore B. cereus spores with high thermal tolerance do exist but appear to be relatively rare. Stoeckel et al. commented that the spores of the B. cereus strain (IP5832) that they evaluated in infant formula were an example of a highly heat-resistant B. cereus spore and that heating processes capable of controlling it could be assumed to inactivate native spore populations in milk products. It had a D100-value of 1.83 min; it was obviously less heat-tolerant than the strains reported by Franklin, Bradshaw, and Dufrenne.

The range of reported z-values for B. cereus spores is 6.7–13.8 °C, with most in the range of 8–11 °C. Hinricks and Atamer cite the z-value for a reference strain as 9.4–9.7 °C, which is in the middle range of reported values.

The above D- and z-value data for B. cereus spores demonstrate a wide range of heat resistance of individual strains. However, some authors have used reported D- and z-values to construct temperature–time semi-log curves of equal destruction of B. cereus spores. For example, constructed lines for a 6-log reduction, while Hinrichs and Atamer constructed a 3-log reduction line. However, based on the information above, the appropriate lines for destruction of individual strains will vary considerably in both position, according to their D-values, and slope, according to their z-values. The wide range of heat tolerances of B. cereus spores makes the construction of representative thermal destruction curves very difficult. Published graphs should therefore be used as a guide to temperature–time combinations to use in processing but should not be assumed to apply to all strains of B. cereus. Interestingly, the D121-value estimated from the graphs of de Jong et al. and van Asselt and Te Giffel is ~0.03 s (that is, 1/6 of the 6-D of ~2 s read from the graphs) and from the graph of Hinrichs and Atamer is ~0.14 s (that is, 1/3 of the 3-D of ~4.2 s read from the graph). These values however, differ considerably from those recorded, above, namely, 0.03–0.04 (1.8–2.4 s) (excluding data for very heat-tolerant strains). Bradshaw et al. isolated a B. cereus strain that had a D121.1-value of 0.03 min (1.8 s) and a z-value of 7.9 °C, and commented that these D- and z-values were similar to those most commonly reported for B. cereus. This approximately agrees with van Asselt and Zweitering: the mean D120-value of B. cereus spores of 0.041 min (2.46 s), (their upper 95% prediction interval D120-value was 0.52 min or 31.2 s) and a D121 reference value given by Hinricks and Atamer, of 0.04 min (2.4 s).

An important factor in the heat resistance of B. cereus spores is their altered behavior when they form biofilms attached to equipment surfaces. Simmonds et al. found an average increase of 205% in D90 values of three B. cereus strains when attached to stainless steel compared with those of planktonic cells. Similarly, Pfeifer and Kessler reported increased heat resistance of B. cereus spores trapped between a silicone-rubber seal and a stainless-steel surface.

- Activation, germination, and growth: Spores of B. cereus have to germinate and the resulting vegetative cells grow before they can cause spoilage or produce toxin. Often spores need to be activated, for example by heat treatment, before they can germinate. However, B. cereus spores are able to germinate without preliminary heat treatment although the rate of germination and the proportion of spores that germinate are higher when the spores are subjected to a heat treatment. B. cereus spores can be activated by heating in milk at temperatures in the range 65–95 °C for various times. The literature varies with regard to the optimum activation conditions, e.g., 65–75 °C, 74 °C for 10 s, 80 °C for 15 s, >80 °C, 85 °C for 2 min, 95 °C for 15 s, and 115 °C for 1 s. The germination medium as well as the temperature is significant. Wilkinson and Davies reported that milk heated at 65 to 75 °C for

15 s provided the best medium for germination, while Stadhouders et al. found that milk heated at 94 °C for 10 s was a better germination and growth medium than HTST-pasteurized milk.

A complication with germination of B. cereus spores is that they exist as both slow-germinating and fast-germinating, with the slow-germinating spores requiring more intense heat-activation treatment than the fast-germinating spores. Heating fast-germinating spores of B. cereus in milk at 65 °C for 2 min or 72 °C for 10 s caused almost total germination at 20 °C in 24 h, while heating the slow-germinating spores at 85–90 °C for 2 min resulted in the same level of germination. It has been shown that a germinant or germination factor is produced in milk by heat treatment. Therefore, pasteurization conditions are sufficient to cause germination of the fast-germinating spores of B. cereus, but not the slow-germinating spores. Fast-germinating spores may be activated by ESL heat treatments although specific reports of this effect have not been located. These fast-germinating spores of B. cereus originate from soil, manure, and fodder, whereas slow-germinating spores seem to come from equipment surfaces. Another issue relating to the spores with differing germination rates is that slow-germinating strains are more heat-resistant than fast-germinating strains.

Temperature has a major effect on the growth of vegetative cells following activation/germination. B. cereus spores may germinate at low temperatures but not show growth at these temperatures for months. The abilities of B. cereus strains to germinate and to grow at low temperatures are not necessarily correlated. For example, Anderson Borge et al. showed that, in a mixture of 11 mesophilic and psychrotolerant strains, the psychrotolerant strains exhibited both the highest and the lowest germination rates in milk at 7 and 10 °C.

- B. cereus in perspective: There is no doubt that B. cereus is a spore-former of some concern to the dairy industry. The above discussion shows that it (or members of the B. cereus group in general) is widespread in the environment and a common contaminant of raw and heat-treated milk. It belongs to a group that contains eight closely related species and hence its precise identification in the reported studies cannot be assured. It has a wide range of growth temperatures amongst its strains; most strains are mesophilic but some are psychrotrophic, which means they can grow in ESL milk during refrigerated storage. The strains vary considerably in heat resistance; most are readily inactivated by common ESL heat processing but some strains are more heat-resistant. Some strains are pathogenic but produce very little toxin at low temperatures. The probability of encountering a strain of B. cereus whose spores are very heat-resistant and that is psychrotrophic and produces toxin at low temperature is small. However, the risk is much greater if the temperature of storage is elevated. The fact that it forms biofilms and can therefore persist on equipment means that it is a constant threat and cannot be ignored by ESL processors.

Post-processing Contamination

The above discussion relates to the keeping quality of ESL milk as affected by the bacteria in the raw milk that are not killed by the heat process. This situation applies if the milk is packaged aseptically and there is no PPC. However, ESL milk is commonly packaged in very clean, but not strictly aseptic, fillers. In this situation, special precautions are taken to minimize the risk of PPC. Such precautions typically include: use of an aseptic tank or an ultra-clean pasteurized milk holding tank equipped with sterile air blanketing; sterilization of empty final packages or packaging material with hydrogen peroxide, with or without UV irradiation; flushing of the filler with sterile air (HEPA-filtered); sterilizing the filler piping and heads with steam at 120–130 °C for 30 min; and spraying an alcohol mist into the filler before filling is commenced. However, such fillers are not completely sealed, are usually not located in a clean room, and the air in the filling zone is not completely sterile. In addition, non-sterile product is being packed in the filler, which can allow biofilm build-up if cleaning is not carried out effectively. This all amounts to the possibility, although small, of periodic contamination of the product. In fact, this is what has been reported by several authors and is the experience of processors.

The reported bacterial contaminants in thermally-produced ESL milk packed under very clean, but non-aseptic, conditions are mostly Gram-positive. This contrasts with the situation in pasteurized milk, where Gram-negative psychrotrophs, principally Pseudomonads, are generally the most prevalent. The common spoilage organisms in commercial ESL milks included the non-spore-formers Rhodococcus, Anquinibacter, Arthrobacter, Microbacterium, Enterococcus, Staphylococcus, Micrococcus, and coryneforms. These bacteria appeared to enter the ESL milk from the air, the equipment, and/or the packaging material. Mugadza and Buys also isolated Gram-positive bacteria from ESL milk, namely Arthrobacter, Annerococcus, and Mycobacterium.

Mugadza and Buys also found the spore-formers B. pumilus (the dominant spore-former), B. subtilus, Paenibacillus, and B. cereus in ESL milk. Paenibacillus and B. cereus, as well as the non-spore-former Micrococcus luteus, were isolated from filler nozzles, which the authors proposed to be a source of contamination of the milk. the common occurrence of B. cereus in ESL milks is of concern because of its potential to be pathogenic and cause spoilage. Mugadza and Buys identified several B. cereus isolates from ESL milk and associated equipment. Based on discriminatory polymerase chain reaction (PCR) analysis, all isolates were shown to have the cspA gene, indicating that they were psychrotrophs. In addition, these isolates produced proteases and hence have spoilage potential. There was a close relationship between B. cereus isolates from filler nozzles and those in milk, indicating the equipment as a source of contamination of the ESL milk. Paenibacillus has been isolated from UHT milk during a period of "tenacious periodical contamination", indicating contamination from processing or packaging equipment.

The above reports strongly suggest the involvement of biofilms in PPC. This is further supported by the fact that the contaminating bacteria in ESL milk have been identified to be processor-specific, which is consistent with, where B. circulans, B. licheniformis, or B. pumilus was found to be the dominant organism in each of three different reports. This is similar to findings for contaminants in pasteurized milk and suggests the need for processor-specific remedial action to reduce bacterial contamination in market milk. It is worth noting here that it has been calculated that one psychrotrophic spore per mL can cause spoilage in 18 d at 4 °C. Hence a very low level of PPC in ESL milk packaged under clean but not aseptic conditions can limit the shelf-life of ESL milk. Unfortunately, such contamination can be sporadic with isolated batches and even some but not all packages in a batch being contaminated; such contamination should be eliminated by aseptic packaging.

Aseptic Packaging of ESL Milk

It is apparent that the shelf-life of thermally-produced UHT milk packaged under non-aseptic conditions is limited by PPC and that it is difficult to prevent PPC in every package in every batch of ESL milk packaged in this way. Therefore, aseptic packaging deserves consideration. While aseptic packaging is usually associated with UHT products, it is recognized that it is highly desirable, though not essential, for ESL products. Aseptic packaging operating correctly eliminates PPC and hence the shelf-life of the product is determined solely by the microorganisms that survive the heat treatment and can grow at low temperature, that is, psychrotrophic spore-formers such as B. circulans, B. cereus, B. pumilus, and Paenbacillus. While the shelf-life of ESL milk packaged aseptically is seldom reported, it has been suggested that a refrigerated shelf-life of 90 days is possible; by comparison, with very clean rather than aseptic filling, a shelf-life of 30–40 days can be expected. Brody cites defect rates for ESL product packaged in ultra-clean and aseptic fillers as 1 in 1000 and 1 in 10,000, respectively. The latter is also the target for UHT milk.

Blake et al. used direct UHT heating at 120–140 °C for 4 s with aseptic packaging and obtained a shelf-life of >60 days as judged by a formal taste panel. Continued sensory testing of the milk samples by a small informal panel found the milks still acceptable after 240 days. It is therefore apparent that very long refrigerated shelf-lives can be achieved with aseptic packaging and, provided the heat treatment is sufficient to inactivate the spores of any psychrotrophic pathogenic spore-formers, such as B. cereus, the safety of the ESL milk can be assured.

Storage Temperature

The temperature of storage of the ESL milk has a major effect on its shelf-life. This was demonstrated by Rysstad and Kolstad, who stored pasteurized milk, which had been aseptically packaged, at 6, 8, and 10 °C. The times taken to reach a total count of 106 CFU/mL were 40, 15, and 7 days, respectively.

A major practical issue for ESL milk is that the temperature of the market cold chain and of consumers' refrigerators cannot be guaranteed to be maintained at this temperature; for this reason, several researchers have carried out storage trials of ESL milk at 7 °C and even up to 10 °C. The only bacteria that should grow in ESL milk are psychrotrophic bacteria, but if the temperature of storage exceeds ~6 °C, some mesophilic bacteria can grow and contribute to spoilage. Such organisms could be the result of PPC or could be from heat-resistant spores surviving the ESL process.

Optimizing the Flavor of ESL Milk

As well as the microbiological aspects, the chemical effect of the heat process also needs to be considered. The main chemical effect of concern is the production of flavor volatiles, which impart a heated or cooked flavor to the milk. The ideal ESL milk should have a flavor similar to that of pasteurized milk, that is, a very low level of heated or cooked flavor. Fortunately this is possible, although in practice the heating conditions for this to occur are seldom optimized.

At this point it is necessary to reiterate an essential difference between the kinetics of bacterial destruction and those of chemical reactions. Heating at high temperatures for short holding times favors high levels of bacterial destruction, while heating at lower temperatures for longer holding times result in high levels of chemical change. This is illustrated in a comparison of UHT processing, which typically occurs at ~140 °C for a few seconds, and in-container sterilization, which uses temperatures of 110–120 °C for 10–20 min. Both processes have a similar bactericidal effect but much more chemical change occurs in the in-container sterilized milk, as evidenced by a brownish color and a distinct cooked flavor. Similarly for ESL processing for a particular bactericidal effect, i.e., the same Fo or B*, combinations of higher temperatures for short times produce less chemical change than combinations of lower temperatures for longer times. This was illustrated by Rysstad and Kolstad with a direct and an indirect process having equal bactericidal effects (i.e., the same B* or Fo) (135 °C for 0.5 s and 127 °C for 1 s, respectively); the indirect process causes much more chemical change, as measured by denaturation of β-Lg, than the direct process.

The chemical effect of a heat process can be assessed in several ways using the kinetics of various chemical reactions. A commonly used chemical index is C*, which is based on the kinetics of destruction of the vitamin thiamine. A C* of 1 is equivalent to a 3% destruction of thiamine and is the recommended upper limit for UHT milk. Another measure is the percentage denaturation of the whey protein, β-Lg. This is arguably a better measure than C* for ESL milk as the denaturation is accompanied by the formation of volatile sulfur compounds, which are largely responsible for the heated or cooked flavor of heat-treated milk.

Mayer et al. investigated the levels of undenatured β-Lg in commercial milk samples, including 71 ESL, obtained from retail outlets in Austria. Only 45% of the ESL milk samples analyzed had β-Lg contents of >1800 mg/L milk, a proposed limit for ESL milk. A further 55% of the analyzed ESL milk samples had low undenatured β-Lg levels, <500 mg/L, equivalent to >85% denaturation). This spread of data reflects the fact that there are no specified conditions for ESL processing.

Rysstad and Kolstad reported 13.6% denaturation of β-Lg in Pure-Lac ESL milk processed at 135 °C for 0.5 s. They compared this with an indirect heating system of equal Fo-value with a nominal temperature–time condition of 127 °C for 1 s, which had 83.5% denaturation. Huijs et al. reported ~20–25% denaturation of β-Lg in milk treated with Innovative Steam Injection (ISI) heating at 150–180 °C for <0.1 s.

Cooked/sulfurous flavor begins to be noticed in heated milks when denaturation of β-Lg reaches ~60%. Therefore, appropriate heating conditions should be chosen in order to minimize denaturation of β-Lg and avoid development of this flavor. Therefore, the recommended level of 1800 mg/L of undenatured β-Lg (45–50% denaturation) is reasonable. A percentage denaturation of 50% is proposed here as the upper limit for ESL milk. It is recognized, however, that only the undenatured level is measured and the level in the raw milk from which the ESL milk is produced is seldom known. For that reason, a level of 1600–1800 mg/mL is suggested, assuming a level of β-Lg in raw milk of 3200–3600 mg/mL. An alternative measure is the undenatured whey protein index (WPNI), which is commonly used to indicate the severity of the pre-heating process in the manufacture of skim milk powder. ESL milk fits into the medium-heat range and it is proposed that the level in ESL milk should be 3.75–4.0 mg/g of dry matter. WPNI correlates very well (negatively) with denaturation of β-Lg.

Another chemical measure of heat load is furosine, which is a measure of the first stage of the Maillard reaction between lactose and protein-bound lysine. Lorenzen et al. reported the level of furosine in commercial thermally processed ESL milk in Germany, Austria, and Switzerland to be 11.1–22.6 mg/100 g protein, while Mayer et al. analyzed 71 ESL milk in Austria and found that only 45% had furosine levels <40 mg/100 g protein, a recommended upper limit. According to these authors, "good" ESL milk had a furosine level of 11.6 mg/100 g protein, while "bad" ESL milk had

71.3 mg/100 g protein. Gallman reported an average furosine level for ESL milk of 20 mg/100 g protein but suggested that it should be possible to achieve <12 mg/100 g protein (and >1800 mg/L of undenatured β-Lg).

Lactulose, which is not present in raw milk, is formed from lactose during heating and is arguably the best chemical index of heat treatment (pp. 198–200). It has been used for ESL milk, with proposed values for good-quality milk of <30 mg/L and <40 mg/L being suggested. For reference, Pure-Lac milks were reported to have <40 mg/L and commercial directly processed UHT milks have levels of ≥90 mg/L.

Optimizing ESL Heating Conditions

Unlike the temperature–time conditions for pasteurization, which are specified in most countries to be ≥72 °C for ≥15 s, there are generally no such specified conditions for ESL processing. The reported temperatures for producing ESL milk by thermal means alone vary from 90 to 145 °C. However, most ESL milk is processed at 120–130 °C, for holding times of a few seconds. In the USA, ESL milk, called "ultra-pasteurized milk," is defined as being processed at ≥138 °C for ≥2 s. Packaging of ESL milk is commonly under ultra-clean conditions but can be aseptic.

Reported commercial processing conditions for ESL milk are mostly in the range 123–127 °C for 1–5 s. Higher temperatures can also be used. As mentioned, U.S. regulations define the process of "ultra-pasteurization" as heating milk at ≥138 °C for ≥2 s. A heat treatment of 138 °C for 2 s may seem severe for ESL milk, but it is still a sub-UHT treatment with calculated F0, B*, and C* of 1.7 min, 0.4, and 0.12, respectively, for a steam infusion or steam injection system; corresponding UHT indices are ≥3 min, ≥1, and ≤1, respectively. Even higher temperatures for shorter times, up to 145 °C for <1 s, using the Pure-Lac system and up to 180 °C for <0.1 s using the ISI system, have also been advocated. These higher-temperature treatments with very short holding times are still sub-UHT treatments in terms of their sporicidal effects, i.e., B* values are <1. ESL milks produced at lower temperatures are also marketed. In South Africa, ESL milk is processed at 94–100 °C. This product is similar to the ESL milks, involving processing by direct steam heating at 89–100 °C.

Several authors have reported the effects of different ESL temperature–time combinations on the shelf-life of the product. For example, Ranjith showed that heating at <117.5 °C for 1 s was insufficient to prevent bacterial growth in cream during storage at 7 °C. He concluded that temperatures ≥120 °C were required for a shelf-life of 49 days at 7 °C. Blake et al. found that treatments at ≤132 °C were insufficient to prevent growth during refrigerated storage but that heating at ≥134 °C for 4 s produced ESL milk with a long shelf-life. In both of these studies, the ESL milk was packaged aseptically.

It is apparent that the optimum heating conditions for producing ESL milk should be based on inactivating spores of psychrotrophic bacteria and minimizing cooked flavor. It is proposed that the former be based on the data of Blake et al., which showed that treatment at ≥134 °C for 4 s was effective. This is in line with the conclusion of Bergere and Cerf that milk processed at ≥134 °C should not contain B. cereus spores. A heat treatment 134 °C for 4 s would have a B* of 0.32 (and F0 of 1.33), considerably less than 1.0, the minimum for UHT processing. This is shown in Figure by "ESL line 1," which joins points with a B* of 0.32 and is to the left of the B* = 1 line. The slope of the line is based on the assumption that the z-value for psychrotrophic spore inactivation is the same as for inactivation of thermophilic spores, i.e., ~10. This is a reasonable assumption given the wide range of z-values reported for B. cereus spores. It is therefore proposed that the B* for ESL processing should be >0.3.

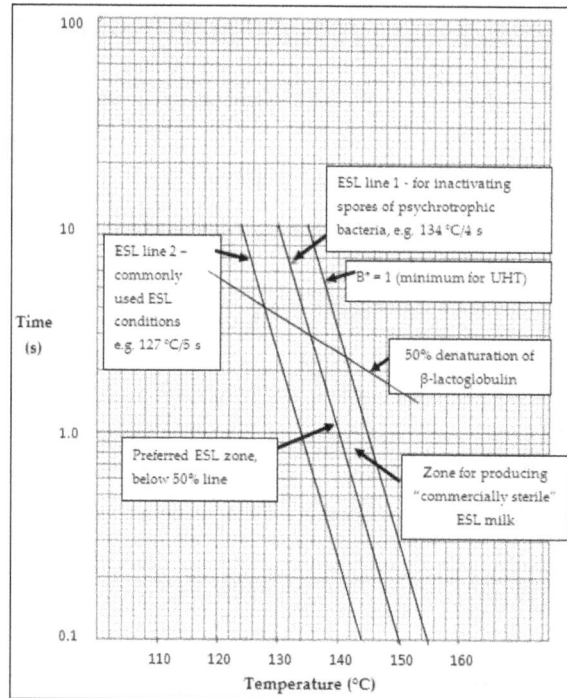

Temperature–time combinations for producing extended shelf life (ESL) and ultra-high-temperature (UHT) processed milk. Points along lines have equal effectiveness in either inactivating bacterial spores or causing 50% denaturation of β-lactoglobulin.

The heat input for ESL milk processing should be less than that which would theoretically cause ~50% denaturation of β-Lg. The temperature time conditions to achieve 50% denaturation of β-Lg are shown by the line labeled thus in figure. Hence, points along the ESL line 1 below the 50% β-Lg denaturation line (the 50% line) represent the best conditions for producing ESL milk to ensure the destruction of spores of psychrotrophic bacteria and cause minimal flavor change. "ESL line 2" in figure shows temperature–time conditions equivalent to 127 for 5 s, which is representative of the conditions commonly used for ESL milk production, and joins points with a B* of ~0.09. Clearly these conditions lie to the left of ESL line 1 and would not be expected to inactivate all psychrotrophic spores.

Points in the zone between ESL lines 1 and 2 and below the 50% line represent reasonable conditions for commercial production of ESL milk; the nearer the conditions are to ESL line 1, the more likely that spores of psychrotrophic spore-formers will be destroyed. Furthermore, points in this zone below the 50% line will have the freshest flavor. Points in the zone between ESL lines 1 and 2 and above the 50% line will have similar bactericidal effects to those below the 50% line but will have more cooked flavors.

Milk processed at temperature–time conditions in the zones between ESL line 1 and the B* = 1 line and below the 50% line will have the greatest bacterial stability, having B* values between 0.32 and 1 and little cooked flavor. In effect, such milk could be termed "commercially sterile" ESL milk. A commercially sterile product is defined as one in which no bacterial growth occurs under the normal conditions of storage; for ESL milk, this is under refrigeration, preferably at ≤4 °C. While this term is normally applied to UHT milk, it is also applicable to ESL milk processed to inactivate all spores of psychrotrophic bacteria, and packaged aseptically. Commercial sterility implies that not all packages of every batch will be devoid of bacteria that could grow and cause spoilage. Brody suggested a target defect rate of ~1 in 10,000, the same as for UHT milk.

Commercially sterile ESL milk with an expected long shelf-life has an increased risk of developing bitterness. The native milk plasmin will not be inactivated under these heating conditions and, although it has low activity at low temperature, it is not inactive. De Jong showed that ISI-ESL milk, a commercially sterile ESL milk that had plasmin activity, did not develop bitterness during storage at 7 °C for up to 28 days. However, bitterness may develop during longer periods of storage. In addition to plasmin, residual bacterial proteases from growth of psychrotrophic bacteria in the raw milk before processing will be more likely to cause proteolysis, and hence bitterness, during long storage times (>30–40 days) than during shorter storage times. For both plasmin and bacterial proteases, maintenance of low temperature in the cold chain, preferably at ≤4 °C, is crucial. Fluctuations to higher temperatures will increase the risk of proteolysis and the development of bitterness. Further research is required to assess the risk of bitterness development in "commercially sterile" ESL milk, with long shelf-life, from proteolysis by plasmin and bacterial proteases.

Assessment of Some Possible Temperature–Time Conditions for ESL Processing

The predicted bactericidal (B*) and chemical (β-Lg denaturation) effects of some possible temperature–time combinations for ESL processing are shown in Table. These have been computed using Excel, as reported by Browning et al. and Tran et al., assuming the process is direct heating with a preheat temperature of 70 °C. The time assumed for reaching the required temperature from the preheat temperature of 70 °C, and also returning to 70 °C after the high-temperature holding, is 0.5 s. This illustrates the low sporicidal effects of 120 °C for 9 s and 127 °C for 5 s and the levels of β-Lg denaturation, which are higher. By contrast, the minimum heating conditions for producing ESL (ultra-pasteurized) milk in USA, 138 °C for 2 s, has a B* of 0.4, which meets the sporicidal criterion proposed here. In Figure, these conditions lie just to the right of ESL line 1. This would theoretically cause ~45% denaturation of β-Lg, below the 50% line in Figure. Therefore, this milk meets the criteria proposed here. The conditions in the last two rows of Table clearly meet the proposed criteria for both B* and β-Lg denaturation and would be expected to produce milk with a fresh flavor and long shelf-life, if packaged aseptically. These conditions are in line with recommendations from some equipment suppliers that produce systems enabling such short holding times to be achieved.

Table: Theoretical bactericidal and chemical effects of possible temperature–time combinations for producing ESL milk.

Heating Conditions (°C/s)	B*	β-Lactoglobulin Denaturation (%)	Comments
120/9	0.03	61	B* too low to inactivate spores of psychrotrophic bacteria; β-Lg denaturation too high
127/5	0.09	55	Representative of commonly used conditions for ESL milk; B* too low to inactivate spores of psychrotrophic bacteria; β-Lg denaturation marginal

Heating Conditions (°C/s)	B*	β-Lactoglobulin Denaturation (%)	Comments
134/4	0.32	56	Conditions sufficient for inactivating spores of psychrotrophic bacteria; β-Lg denaturation marginal
138/2	0.40	45	Minimum conditions for ESL in USA; meets proposed criteria
140/1	0.32	34	Meets proposed criteria; excellent conditions if short holding time can be achieved
145/0.3	0.32	24	Meets proposed criteria; excellent conditions if short holding time can be achieved

Temperature–time lines for 6-log reduction and 3-log reduction of B. cereus spores have been published. If these were placed on Figure they would be far to the left of ESL line 2. The first passes through the point 130 °C for 0.1 s and the second passes through the point 130 °C for 1 s. The conditions along those lines would clearly not meet the criteria for ESL milk proposed here. The approximate B* values would be 0.01 and 0.04, respectively.

It is also instructive to indicate the position on Figure for the temperature–time line of inactivation (9-log) of spores of mesophilic spore-formers. Such a line was published by Kessler and reproduced by Chavan et al. It would be approximately halfway between ESL line 1 and the B* = 1 line. It would pass through the point of 135 °C for ~8 s and have a B* of ~0.8. This is clearly in excess of what is proposed here for ESL milk but, provided the temperature–time conditions fell below the 50% line in Figure, they would produce high-quality ESL milk. As the B* would be <1, milk processed with these conditions would not meet the recommended bactericidal criterion for UHT milk.

In several countries, the specified temperature–time conditions for UHT processing are ≥135 °C for ≥1 s. Therefore, milk processed at the lower end of the ranges is effectively ESL milk. A direct thermal process operating at 135 °C for 1 s has a F0 of 0.45 and a B* of 0.11, which are considerably less than the minima required for a UHT process, i.e., F0 = 3 and B* = 1. However, if the ESL milk processing conditions fall within the UHT range (≥135 °C for ≥1 s.), the ESL milk may not be able to be labeled "fresh," a practice ESL milk producers prefer to follow.

References

- Thermal-destruction-microorganisms, foodscience: uoguelph.ca, Retrieved 31 March, 2019

- Hickey, D.K.; Kilcawley, K.N.; Beresford, T.P.; Wilkinson, M.G. (2007). "Lipolysis in Cheddar Cheese Made from Raw, Thermized, and Pasteurized Milks". Journal of Dairy Science. 90 (1): 47–56. Doi:10.3168/jds. S0022-0302(07)72607-3. PMID 17183074

- Spray-drying-of-dairy-products-a-review: newfoodmagazine.com, Retrieved 9 May, 2019

- Fletcher, Janet (July 13, 2006). "Wisconsin's Buttermilk Blue adds pizazz to summer salads". San Francisco Chronicle. Hearst Communications Inc. Retrieved October 23, 2010

- Sterilization: dairyknowledge.in, Retrieved 8 August , 2019

Technologies used in Dairy Industry

CHAPTER 5

Various technologies are used in dairy industry such as separators, homogenizers and packing machines for milk. There are also various cheese making, butter making and ice-cream making equipment that are used in the dairy industry. This chapter discusses in detail these technologies and equipment related to dairy industry.

CHEESE MAKING EQUIPMENT

Cheese is protein rich food. Cheese is a concentrated milk product of great importance. It commonly includes the steps:

- Setting the milk to develop some acidity.
- Coagulation by an enzyme to facilitate cutting.
- Cutting the coagulum for easy expulsion of whey.
- Cooking or heat treatment to shrink the curd and drain the whey.
- Consolidation and shaping.
- Curing or maturing the green cheese for short or long periods of time.

The traditional process of cheese making consists of:

- Standardization, clarification and pasteurization of milk to get uniform quality of cheese.
- Homogenization of milk to help in greater access of lipase enzyme to fat, allowing faster fat hydrolysis necessary to secure blue cheese flavour, wherever necessary.
- Formation of coagulum involving use of favourable starter culture (0.5 to 1.0%) at 30 to 31 °C for development of acidity (0.01 to 0.02%) and to assist curd formation by rennet added at the rate of 165 ml/1000 kg milk in about 30 min.
- Careful cutting of the curd into uniform size to promote whey removal.
- Raising temperature of curd to 35 °C in about 30 min and holding for about 45 min to firm the curd and removal of whey.
- Dipping or whey drainage and retaining maximum solids.

- Cheddaring, a characteristic step for cheddar cheese, by cutting the curd into slabs, turning every 15 min and piling every 30 min to get 3 to 4 high piles of curd to develop proper body in the curd when acidity of whey is 0.5% lactic acid (pH about 5.2).

- Milling the curd into pieces to help in filling into hoops for pressing.

- Salting to get about 1.5% of salt in finished cheese for flavour development and stoppage of acid production.

- Pressing hoops for about 12 to 16 h for removal of whey and getting final desired moisture content in cheese, and cheese blocks waxed to cover external surfaces.

- Curing under controlled conditions of temperature and humidity to develop characteristic body and flavour in cheese.

Equipment

Cheese equipment consists of cheese vats, cheese knives, agitator, curd strainers, curd mill, cheese hoop and press.

Cheese Vat

The cheese vat used for coagulation and cooking are made of either SS or with SS lining rectangular design. The vats are jacketed, allowing space for circulation of water between the inner and outer container. A steam pipe is fitted at the bottom in between outer and inner tank for uniform distribution of heat to the inner tank. The outer tank may be MS construction. The inner tank should not have any sharp corners and the welded joints should be properly ground and polished for effective cleaning. Fittings such as jacket drain valve, overflow connection, SS gate valve to draw whey with longitudinal central ridge to allow whey to flow towards whey valve with steam and water inlet connections are fixed on the vat. The vats are supported suitable on adjustable legs.

Agitator

The agitator is used for moving the curd after cutting, fork type motor driven agitator which reciprocates back and forth from one end of the vat to the other end is generally used in large vats. The stroke of this agitator is adjustable between two stops. A paddle type agitator or wooden rake is used in small units.

Cheese Knives

Uniformity in the size of the pieces is the aim of proper cutting. For this purpose, two kinds of knives are used. One knife cuts the curd into horizontal layers and the other vertically across these layers from top to bottom, slicing them into small cubes. The blades are thin wire which gives effective cutting. In some the blades are thin and sharp of metal construction, cuts the curd with least possible breaking. The distance between the blades usually varies from 6 to 17 mm.

The curd is cut first with horizontal knife lengthwise of the vat, then crosswise with vertical knife and finally lengthwise with the same knife. The knives may be of SS construction or tinned brass.

Gate Strainer: Whey is removed from the vat through a gate strainer which holds back the curd. It is semicircular in design with perforation which can be fitted to the vat outlet.

Curd mill: The mill, spike-toothed or circular blade type cuts the curd into small pieces of uniform size and should do it without crushing or squeezing the milk fat from the curd. The mill could be hand operated or motor driven. If power curd mill is used, it should not be run too rapidly to prevent uneven curd cutting which will result in poor cheese texture.

Cheese hoop: Different designs and capacities of hoops are available, to give shape to the curd and compact the same. They are round or rectangular (Wilson type) with followers.

Cheese press: The cheese press is used to press the cheese in the hoops. The loaded cheese hoops are placed in the press and pressure applied. The hand operated cheese press consists of a frame having two or three vertical columns, pressing plates fitted to sleeves moving on the vertical columns, a simple or compound lever attachment or hand wheel with spindles and proportional weights for applying pressure. The press could be vertical or horizontal, mechanical or hydraulic type depending upon whether the force is applied by mechanical action or hydraulic pressure. The pressure on the cheese should be uniform and there should not be any buckling of the hoops during pressing.

Mechanization in Cheese Making

Mechanization is a system in which most of the stages in cheese making are carried out by machinery instead of manual labour. The present day cheese making systems involve handling milk ranging from 2,50,000 to 10,00,000 l/day. With increasing cost of labour and conventional horizontal vats it is very difficult for cheese makers to cut the coagulum and handle the curd as required for the conventional cheddar process. When the volume of milk increases it is logical to reduce manual work and drudgery. The adaptation of mechanical principles in cheese making has grown with the trend of automation in dairy industry. Mechanization in the major stages of making cheese is given below.

Curd making

Use of banks of rectangular jacketed cheese vats of about 15000 l capacity using built in swinging stirring units to stir curd-whey mixture after cutting coagulum manually is an accepted practice. Mechanical cutting is now widely used and nylon thread is replacing SS wire in many cases. Vats are fitted with pneumatic or hydraulic tilting gears for emptying the contents.

Besides horizontal enclosed cheese processing tanks, totally enclosed round ended vertical cheese vats with built-in-stirring units usually made of two frame assemblies with angle of the blade in the knife agitator frame is so designed that the assembly acts as cutting knife in one direction and as agitator in the opposite direction. The knives have rotary or linear movement depending on the type of vats. The vertical vats economise space. In enclosed vats many operations like filling, adding starter, renneting, cutting, stirring, scalding and emptying are controlled automatically.

Curd/Whey Separation and Texturing of Curd

For the separation of curd pieces from whey various devices are used, viz., troughs with perforated linings, vibrating mesh separator to separate fine particles of curd from whey, rotating screens or simple screens.

The curd conditioning is done by (a) pumping or gravity flow of curd/whey mixture, (b) transfer of curd pieces by vibrating belt or conveyor belt. The stretching and squeezing are affected by belts running at different speeds. The curd is converted into fibrous slab in cheddar towers where the curd is forced down by the pressure of the succeeding curd into a square bottom cross section, or in a cheddar box with regular turning through 900 every 15 min. milled, salted and filled in moulds and carried on long moving belts.

Hooping, Moulding and Pressing

The milling of cheddared curd is done by power driven mill and the use of block shaped perforated SS or light non-corrodible plastic moulds. The feeding of the milled curd is done by vibrations, and automatic weighing machine.

The pneumatic, hydraulic or mechanical spring press has replaced the lever type of presses. The demand of close textured cheese has led to the technique of vacuum pressing of cheese. This system sucks out air between the curd particles and eliminates open texture in cheese. The press consists of a chamber of rectangular cross-section fitted with piston at the bottom for compression and extrusion of the pressed curd at the top. The top removable lid can be replaced by cutters to get blocks. The height of the extruded curd is adjusted to get required size of the block. Vacuum is applied on all the sides of the curd block (about 80 kPa). This is followed after about 10 min, a pre-pressing pressure of about 3,400 kPa for 10 min, raised to about 9,500 kPa leads to a final pressure of 480 kPa on the cheese. The press time varies. Normally, it is 1.5 h under vacuum and 6.5 h under normal pressure.

For brine salting multitier crates are used for lifting, immersing and taking back of small cheese blocks from brine bath. For retail use cheese blocks are cut by using mechanical cutters from variously shaped blocks and film packed with suitable materials. Several mechanical cheese making systems are available for cheddar, hard and semi-hard varieties and soft cheese.

Continuous Cheese Making

In this system milk is fed at one end of the machine and continuously converted to coagulum, curd and cheese during the passage through the machine. Basically the process consists of application of heat to cold renneted milk in such a manner that curd is formed continuously and then cut for removal of whey and finally ending with cheese.

BUTTER MAKING MACHINES

Generally the extra fat of milk is best preserved by converting it either into butter or ghee.

Batch Butter Churns

Rotating Churns

The rotating butter churn was introduced in the nineteenth century and gradually from farm butter making it was adopted for the factory butter making by the butter industry. The rotating churns consisted mainly of a barrel rotated on an axis with shelves of various kinds to increase the agitation effect. The first combined churn and butter worker was introduced in USA in 1890. The combined churn and butter worker was of short barrel type.

Batch Method using Rotating Churns

The use of batch churn for butter manufacture is on decreasing trend because of increase in popularity of improved designs of continuous butter making machines. the butter produced by organized dairies is made by batch churns, except a few leading dairies.

The capacity of batch churns varies from 100 to 3000 Kg of cream per batch. The shape is mostly cylindrical with front opening , cone with cylinder, single cones and double cone etc. The churn is short in length and large in diameter. Baffles are fitted internally to improve agitation. In some designs, ribbed rollers are fitted through which butter grains pass. The fittings like air vent, sight glass, butter milk outlet, opening for cream inlet, and outlet for butter are mounted on barrel. Butter does not adhere on the wood, while the metal churns, the inside surface is roughened (sand or lead shot blasting) to allow film of moisture on the surface between the metal and the butter.

The degree of mixing depends on the amount of cream in the churn and on the rate of revolution. Too low a rate will not give sufficient turbulence and with too high a speed there is the danger that the centrifugal force (m ω2R) will exceed the gravitational force (mg) and that the cream will stick to the periphery and rotate there with drum. The best condition for churning i.e. maximum turbulence, are achieved when the force of gravity just exceeds the centrifugal force.

i.e. $m\omega^2R < mg$

$(2\pi n)^2 R < g$

Or $n < [g/R]^{1/2} . 1/2\pi \approx 1/(2\sqrt{R})$

The energy consumption is about 7 - 11 kWh per 1000 kg of butter of which about 90% is used in churning and 10% in working. The lower values are for the cream with a higher fat content.

Loading the Cream

Pasteurized cream with 35 - 40 percent fat, properly aged is pumped into the churn. Cream is filled to 40 - 45% of the volume of the churn. The cream may be ripened.

Churning

The churn can be operated at different speeds. The range of speed depends on the size and shape of the churn. The cream is churned at the churning speed (60 - 100 rpm). The cream is well whipped by the corners, edges and other irregularities in the churn. Chilled water is sprayed over the churn during churning operation. It takes about 35 – 40 minutes for the formation of butter granules of peanut size.

Buttermilk Draining

The churn is stopped and buttermilk is drained off. Equal quantity of pasteurized wash water is added.

Washing

The churn is started again. The wash water is drained off after some time. Two or three washings are generally given.

Working

The wash water is drained off and salt is added. The churn is then operated at lower speed (25 - 50 rpm) for working as compared to that at churning. After 3 – 5 min., sample is taken and moisture is adjusted by adding required quantity of water. The working is carried out till desired body and texture is attained. Applying vacuum of 5 m of water gauge during working gives close texture by reducing the content of air.

Unloading and Packing

The butter is unloaded in trolleys and then packed for sale. Different types of packing machines are employed for the required size of packages.

Care of Churns

- Driving gear should be filled with lubricating oil and every alternate year replace it.
- Never change the speed while the churn is running.
- Solid foundation is necessary.
- Gaskets to be maintained leak proof.
- Proper roughness inside of the churn should be maintained.
- Proper cleaning of the churn after the operation is over.

Continuous Churns

Continuous butter making was first introduced in 1889 following successful development of centrifugal separation of cream from milk by a continuous process. The machine developed was first exhibited in England in 1889. The machine was known as butter extractor. The separated cream was beaten with great violence and thus converted into butter granules, which was discharged along with the buttermilk.

Continuous Butter Making

Different continuous butter making machines which are used in dairy industry can be classified into three categories according to the churning principle involved as under:

- Machines operating on Fritz process or floatation churning, where accelerated churning and working takes place. Few examples of this type are known as Westfalia Separator (West Germany), Contimab (France), and Masek (Czechoslovakia).

- Machines operating on concentration of normal cream followed by phase inversion, cooling and mechanical treatment. Machines with commercial names as Alfa (West Germany) and Alfa-Laval (Sweden) fall under this category.

- Machines operating on concentration of normal cream de-emulsification and re-emulsification into butter are available with commercial names of Gold Flow Process (USA), New Way Process (Australia) and Creamery Package Process (USA).

The Fritz Process

The first prototype of modern machine developed by Fritz was demonstrated in the year 1940. Only this process has managed to consolidate its place in the Western Europe. This is probably because its close similarity with ordinary batch method and the ease with which it can be applied practically. It contains three main parts, viz. Churning cylinder, draining and washing cylinder and worker. Capacity is up to 10000 kg/h.

Preparation of Cream

For smooth operation of the churn, the cream must be of uniform quality. The properties like fat content, temperature, age, pH, previous treatments it has undergone and like are controlled.

Churning the Cream

The cream delivered by a cream pump, enters the butter shaft through cream inlet. From this, it goes into the churning cylinder. The butter shaft is driven by a variable speed V-belt drive. In this cylinder, is a four-armed beater running at 250 – 2800 rpm with a wall clearance of 2-3mm. The speed can be varied as required. The pockets in the churning cylinder impart extra turbulence to the film of cream thereby enhancing the butter making action and improving the yield. Butter granules are formed within a 3-5 seconds. The cylinder is cooled by cold water circulation during the operation. Second churning cylinder rotates at 10-25 rpm or stationary cylinder with paddle-bearing shafts rotates at 34 rpm it is also cooled.

Butter Milk Separation and Draining

The mixture of buttermilk and butter granules proceed into the adjoining section known as separating cylinder with its welded in ribs. The next unit is the buttermilk draining cylinder. This cylinder is perforated and provided with beater studs. Buttermilk gets separated and drained through holes.

Washing the Granules

The butter granules enter the washing section which is also perforated and has provision for washing spray. The grains which by now reach the finished stage are washed here.

Working and Texturizing

The washed butter grains fall into twin worm butter worker (two contra rotating screws). They are inclined and forces butter through number of perforated plates arranged in series which gives the fine dispersion of water. The process is assisted by mixing vanes, situated between plates. The butter worker can be operated at either of two fixed speed of approximately 65 and 30 rpm. there are provisions for salt and colour addition, moisture corrections etc. The butter comes out in the form of a continuous stream, its shape depending upon that of the outlet spout.

Concentration and Phase Inversion Process

Machines operating on the principle of concentration and phase inversion were introduced after the discovery of Wendt of high fat cream. He found that normal cream could be re separated to a rich cream with fat content as high as the minimum permissible fat content in butter. Butter formation in the cooling of concentrated cream was observed by Mohr. The Alfa process was developed in Germany and Sweden. The Alfa-Laval (Sweden) butter making process consists of the following major steps.

Pasteurization and Concentration

Normal cream of 30 – 35 percent fat is pasteurized at 90 °C. It then passes through a cream concentrator where the fat content is raised to 80 – 84 percent.

Cooling and Phase Inversion

The high fat content cream is delivered by the concentrator into the balance tank, which is mechanically stirred continuously. The prepared cream is then drawn and is forced through the transmutator. This consists of a bank of three stainless steel jacketed cylinder provided with mechanically driven rotors. There is an annular space about 0.5 to 0.6 cm wide between the inner cylinder and the rotor. The rotors are fitted with soft metal ribs set in spiral fashion and of such thickness as to scrape lightly the internal surface of the cylinder. The direction of rotation of the rotors is related to the direction of spiral ribbing so as to aid the forward movement of the contents in the annular space. Brine is circulated through the jackets. The cream is kept moving over the cooled surfaces of the cylinders by the revolving rotors. The combined cooling and mechanical action causes butter formation to take place. The butter leaves the transmutator in a semi liquid state, but solidifies rapidly. A plate type transmutator is now available.

The Alfa process is not suitable for acid cream as it clogs the concentrator bowl. It involves some difficulties in control of moisture content of butter.

Concentration, De-emulsification and Re-emulsification Process

The main steps of Gold'n Flow butter making method are described below.

Destabilization

Freshly separated cream of 30 – 40 percent is taken and air is incorporated in it. After this, the cream passes through a destabilizing pump containing perforated rotor blades turning at about 3000 rpm. Better destabilization is obtained by using two pumps in series. The cream is preheated to 60 °C in a vertical centrifugal heater. The preheating is employed to bring the cream to adequate temperature and to complete destabilization. Fat globule membrane is disrupted and weakened by the pump and the destabilization is completed by subsequent heating. Fat then moves as a continuous mass which is free of globular fat.

Concentration

After destabilization cream is concentrated to 85 – 90 percent by a special separator which discharges skim milk, sludge and concentrated cream.

Pasteurization

The concentrated cream is pasteurized in a vacreator at 88 - 90 °C and subsequently cooled to 38 – 43 °C.

Standardization

The liquid butter fat in which small droplets of cream are dispersed is constantly stirred and standardized for moisture, salt and fat.

De-emulsification and Texturisation

The standardized mixture is cooled to 4 - 6 °C in a chiller. The chiller consists of two horizontal cylinders, installed side by side, and cooled by direct expansion of NH_3. Each cylinder is equipped with an agitator provided with scrapper blades. The moisture is finely disturbed by means of vigorous mechanical treatment.

The required consistency is obtained in the butter by treatment in the texturator. While passing through the texturator, temperature increases by about 2 °C, which indicates further crystallization of butterfat. Butter comes out at 4 - 6 °C.

SEPARATOR

Manual separator in a Swedish museum.

A separator is a centrifugal device that separates milk into cream and skimmed milk. Separation was commonly performed on farms in the past. Most farmers milked a few cows, usually by hand, and separated milk. Some of the skimmed milk was consumed while the rest was used to feed calves and pigs. Enough cream was saved to make butter, and the excess was sold.

Today, milk is separated in industrial dairies. Sufficient cream is returned to the skimmed milk before sale.

Before the advent of centrifugal separators, separation was performed by letting milk sit in a container until the cream floated to the top and could be skimmed off by hand. A variant container-separator had a nozzle at the bottom which was opened to allow the milk to drain off. A window in the side, near the nozzle at the bottom, allowed the operator to observe when the milk was drained.

The centrifugal separator was first manufactured by Gustaf de Laval, making it possible to separate cream from milk faster and more easily, without having to let the milk sit for a time, and risk it turning sour. Possibly because Gustaf de Laval manufactured the first cream separators, many people credit the invention to de Laval. However, many patents appear before his, all of them labelled as improvements. One of the first specifically for cream separation was patented by W. C. L. Lefeldt and C. G. O. Lentsch.

Mechanism

Manual rotation of the separator handle turns a worm gear mechanism which causes the separator bowl to spin at thousands of revolutions per minute.

When spun, the heavier milk is pulled outward against the walls of the separator and the cream, which is lighter, collects in the middle. The cream and milk then flow out of separate spouts. After separation, the cream and skimming milk are mixed together in a certain ratio until the favoured fat content has been set. The ratio is depending on the product which is to be produced (low-fat milk, fullfat milk or cream). Some floor model separators were built with a swinging platform attached to the stand. The bucket for collecting the cream was put on the platform, and a much larger bucket was set on the floor to collect the milk. Some floor model separators had two swinging platforms. Smaller versions of separators were called table-top models, for small dairies with only a few cows or goats.

Gustaf de Laval's construction made it possible to start the largest separator factory in the world, Alfa Laval AB. The milk separator represented a significant industrial breakthrough in Sweden. Within the first decade of the 1900s, there were over twenty separator manufacturers in Stockholm. Separators in modified form are also used on ships to purify oil, which may have been their original use, because in its original form de Laval proposed the separator for use in his steam turbine. De Laval's turbine used mechanically lubricated journal bearings which weren't insulated from the inside of the turbine. When the steam condensed into water it contaminated the oil. To purify the oil a centrifugal separator was used, which was later adapted to the dairy industry.

The original design had a manual bowl that required manual cleaning. Most modern separators use a self-ejecting centrifuge bowl that can automatically discharge any sedimentary solids that may be present, and that allow for clean-in-place (CIP).

A distinction is made between warm milk skimming and cold milk skimming:

- Warm milk skimming separator: At first the raw milk is heated and then skimmed warm. There is a significant difference in density between cream and skimmed milk, because of the higher temperature.

- Cold milk skimming separator: Because of the lower energy, which is used, the production cost will be reduced. Also at cold temperatures, the growth of microorganisms is significantly reduced. In the USA, Mexico, Australia and New Zealand, cold milk skimming is on the rise.

ICE-CREAM MAKING EQUIPMENT

Ice cream is a frozen dairy product. In ice cream making refrigeration system and air incorporation system is required. The type of freezer play an important role in freezing of ice cream mix and subsequently on the quality of the product. There are batch and continuous type of ice cream freezers available in the market.

Ice Cream Freezer

The function of the freezer are:

- To freeze a portion of the water of the mix to get a smooth product,

- To incorporate a predetermined amount of air uniformly into the mix to get proper overrun.

Fast freezing is essential for a smooth product because ice crystals that are formed quickly are smaller than those formed slowly. Therefore, it is desirable to freeze and draw from the freezer in as short a time as possible.

Failure to provide adequate refrigeration during freezing or hardening, results in formation of large ice-crystals in ice cream. Also, since freezing continues after the ice cream is placed in the hardening rooms; the ice crystals formed during the hardening period are larger because they form more slowly than in the freezer. For this reason, it is desirable to freeze the ice cream as stiff as possible and yet have it liquid enough to draw out of the freezer. Following factors influencing the freezing time,

Mechanical factors:

- Type and make of freezer,

- Condition of the freezer wall and blades,

- Speed of the dasher,

- Temperature and flow rate of the refrigerant,

- Overrun desired and,

- Feed rate.

Characteristics of the mix influencing the freezing time are:

- Composition of mix,

- Freezing point of mix and method of processing the mix.

Freezers are of two basic types: the batch and continuous freezers.

Batch Ice-Cream Freezer

Tub Freezer

It has a vertical can, surrounded by ice & salt mixture. The can is rotated with mix filled to ½ and rotated with handle and bevel gear. The dasher is stationary, while the can will be rotating when handle is rotated. The salt keeps the ice melting point low. This type of batch freezer is now not in common use.

Horizontal Batch Freezer

It has refrigerated drum or cylinder which is stationary, with direct expansion of refrigerant in the coils surrounding the cylinder. The cylinder itself can be of SS or Brass with Nickel coating, to take advantage of good thermal conductivity of the Copper alloy, at the same time to prevent it from coming in contact with the product.

The cylinder has a front openable door, which in its inner side has bush arrangement to hold a horizontal shaft. The door even when closed, the cylinder has an opening at the top to pour ice cream mix, and another opening at bottom to draw frozen ice cream.

Dasher which is a combination of Beater and Scrapper has the following functions:

- To scrape the frozen film from cylinder wall, so that ice crystals do not grow beyond 40 μ.

- Beat the mix to incorporate air to the maximum extent possible.

- Mix any fruit & flavouring material uniformly.

- Eject the finished ice cream rapidly.

- It is important to have the dasher in proper alignment and the blades must be sharp.

- The beater and scrapper rotate in opposite direction to avoid slug type movement of the entire semifrozen mix.

- The Dasher rpm is usually 150 to 250.

The sequence of operations involves:

- Fill the mixture one third to half the volume of the cylinder.

- Start freezing the mix , with switching ON both the refrigeration system and beater.

- Time to time check the stage of freezing and over run by taking the sample and weighting, as well as keeping a watch on the Ammeter reading of the Dasher motor. The power consumed will increase with the progress of freezing.

- After required freezing is achieved (approx. half the moisture to be in frozen condition), continue beating to incorporate air further till required over run is achieved (which is maximum 100% for batch freezer).

- The freezing is generally over in 5 to 7 minutes, and over run takes little more time, if the composition is suitable.

Continuous Ice Cream Freezer

Advantages

- Large scale production capacity.
- uniform and smoother ice cream quality.
- More efficient controls.
- Shorter aging time is possible.
- Less tendency towards sandiness.

Disadvantages

- High initial cost.
- Not suitable for small scale production.

Construction

- The volume of the cylinder is comparatively reduced, by having more solid dasher and mix passing as a thinner layer, as compared to batch process.
- The freezing chamber is supplied with mix from tank by two pumps in some designs, where the first pump is a metering pump, designed to feed mix at a controlled rate. The outlet of this pump leads to a second pump through a pipe, which has provision to allow controlled quantity of air. The second pump being larger in capacity (approx. three times), it handles both incoming air and the mix. The second pump, then, leads the mixture of air and ice cream mix, in to the freezing chamber at about 5 to 6 kg/ cm².
- Other designs single large pump with air entry on suction like along with mix.
- Both horizontal and vertical models are available.
- In a continuous freezer, about 30 second time is taken for partial freezing the mix.
- The ice crystal size is reduced to 45-55 μm and the air cell wall thickness to 100-150 μm.
- For cooling, evaporating coolants in the temperature range of -20 to -30 °C are used in the cooling jacket.

Refrigeration Unit

- Cooling is by Direct expansion using R502 refrigerant (Cattabriga).
- Semihermatic compressor, with oil sum heating element is provided.
- Condenser is water cooled, along with a safety valve in case of high pressure being developed.
- Cooling water is regulated.

- Liquid refrigerant line with provided with Ceramic filter, sight glass, solenoid valve.

- The expansion valve is Thermostatic type.

- Two hours after oil heater is switched ON, the compressor is to be switched on. This removes the refrigerant in the oil.

- Controls include switches for Beater, compressor. When the beater motor current consumption exceeds a limit, the controls activate a solenoid, which allows hot gas from compressor immediately below the expansion valve. This control can be operated manually also. This limiting value can be adjusted by a Rheostat.

Pumps

Pumps of ice-cream freezers are usually of the rotary type with the capability to pump against pressure of 7-14 kg/cm2(690-1380 kPa) with reasonable volumetric efficiency. There are two general pumping arrangements, both designed as a part of the overrun system. The first employ a pump (or a pair of pumps or compound pump) to pump or meter the mix into the freezing cylinder, plus a hold-back valve at the ice cream discharge port. The hold-back valve may be spring loaded with manual adjustment, it may have an air operator with adjustable air pressure supplying the operating power. The hold-back valve permits imposing a pressure on the cylinder during freezing which compresses the air admitted with the mix for overrun. Cylinder pressure of 3.5-4.0 atmospheres keeps the volume of air in the freezing cylinder sufficiently small so that it does not significantly lower the internal heat transfer out from and through the mix. That pressure is sufficient for proper air dispersion and small air cell size. Higher pressures may be imposed on the cylinder, but in most cases, the improvement of heat transfer and air cell size is not great enough to offset the disadvantages of increased pumping cost.

Controls and Automation

All continuous ice cream freezers have control for operation which include on-off switches for pump and dasher motors, and for air compressor motors (when these are part of the freezer), for solenoid valves on hot gas defrost lines, air lines and refrigerant supply lines, speed regulation of pumps, refrigeration supply and back pressure, pressure gauges for the refrigeration system and cylinder or air pressure and dasher motor ammeter, wattmeter or motor load indicator. In addition more sophisticated machines may have a viscosity meter and controller, and a programmable controller or micro-processor to operate and control most functions of the ice cream freezer.

The modern ice cream freezer consists of a micro-processor programmed to control all the function of operation including overrun, viscosity of product, cylinder pressure, all operating steps such as start up, routine or emergency shutdown, resumption of operation after an automatic shutdown when the reason for shutdown has been corrected. Preparation for cleaning and the valve and pump by pass is required for automatic cleaning. The micro-processor shows the time of day, mix flow rate, percentage of overrun, product rate, hours of operation, accumulated production in that time interval, the program step in operation, and various warnings. In case of an impending freeze-up, the warning is displayed and corrective action is taken. If a freeze-up should occur, the micro processor automatically causes defrosting of the cylinder and operation to be resumed when conditions are satisfactory. The display can be in one or more of several common languages.

The micro-processor programmed operation assures that all functions are performed in the proper sequences, and under the conditions envisioned by the designer of the freezer. This is especially beneficial to the ice-cream maker in preventing damage to the freezer in emergency situations, thus avoiding the incidental unplanned down time in production.

Adding Ingredients and Flavours

Flavouring materials are added after the mix has been made. Pieces of fruit and purees should not be added to the mix prior to freezing in continous freezers, as they tend to settle out in the tank with subsequent poor distribution in the frozen ice cream. The ingredient feeders often referred to as fruit feeders have a hopper for the ingredient, an auger or other means for metering or proportioning the fruit, a rotator or plunger for inserting the ingredient.

HOMOGENIZER

The intensive mixing process has existed for decades, beginning with the advent of the blender by Stephen Poplawski in 1922. The equipment has since evolved to something more complex and with more potential functions; users now have the option of a basic blender for simple kitchen processes, or an intensive mixing homogenizer designed for laboratory use. Yet while the general public is familiar with the functions and appropriate occasions for blender use, the same is not true for a homogenizer.

Homogenization works by forcing the sample through a narrow space. Multiple forces (including turbulence and cavitation) in addition to high pressure, can act on the sample to create a high quality product. Specifically because of its powerful pressure, homogenization is frequently used for particle size reduction or cell lysis. These processes can result in a valuable yield, from creating cell lysates containing valuable proteins or DNA to producing emulsions, dispersions, and suspensions. Although a myriad of other mixing machines exist, the homogenizer is superior for multiple reasons. Not only is it easily scalable, but it also uses multiple mechanical forces (as opposed to one, which is the case for most mixing equipment); this results in a stable, uniform, and consistent product.

Multiple industries can, and do, take advantage of the impressive homogenizer capabilities. In the pharmaceutical industry, researchers use the scalable homogenizer first during laboratory research, and then in clinical trials and manufacturing. Its yields are typically incorporated into critical products like vaccines, antibiotics, cancer treatment, and pharmaceutical creams. In the beverage industry, homogenized milk has a longer shelf life and is more stable and physically attractive. And in the biotech industry, particle size reduction can yield nanoparticles and other new innovations.

BASIC PRINCIPLES OF EVAPORATORS

Evaporation and vapouration are two processes in which simultaneous heat and mass transfer process occurs resulting into separation of vapour from a solution. Evaporation and vapourization occur where molecules obtain enough energy to escape as vapour from a solution. The rate of escape of the surface molecules depends primarily upon the temperature of the liquid, the temperature of the surroundings, the pressure above the liquid, surface area and rate of heat propagation to product.

Vapourization and Evaporation

Evaporation and vaporization are quite different from each other.

Evaporation and vapourization occur where molecules obtain enough energy to escape as vapour from a solution. The rate of escape of the surface molecules depends primarily upon the temperature of the liquid, the temperature of the surroundings, the pressure above the liquid, surface area and rate of heat propagation to product. In a closed container with air space above the liquid, evaporation will continue until the air is saturated with water molecules. Removal of water from a liquid product by evaporation is enhanced by adding heat and by removing the saturated air from above the liquid. This is done by removal of vapour from the space above the liquid surface and there by creating vacuum. The boiling point of solution due to dissolved solutes is higher than that of pure water and depends on the molecular weight of the solute. Vacuum is utilized to remove water from liquid/solids at lower temperatures to reduce damage to heat sensitive products which might decompose at higher temperatures.

In the dairy industry evaporation means the concentration of liquid milk products containing dissolved, emulsified or suspended constituents. During this process water is removed by boiling. This process is used in the dairy industry for manufacture of evaporated milk, condensed milk.

In milk condensing plant, milk is condensed by evaporating a part of its water content by using saturated steam. The milk is boiled under vacuum. As the milk boils, water vapour is formed. This vapour is utilized for heating the milk further in the next stage which is at a higher vacuum.

Modern dairy plants use evaporators to remove part of water from milk by boiling it under low pressure. The process of evaporation takes place at a maximum temperature of about 70 oC corresponding to an absolute pressure of 230 mm (9.0 inch) of mercury (Hg). Evaporation of milk under low pressure or vacuum is carried out in a specially designed plant. The plant design depends much on the characteristics of liquid milk during boiling at low pressure than any other factor. Some of the important properties of evaporating milk are as under.

- Concentration of solids (initial and final).

- Foaming under vacuum.

- Heat sensitivity.

- Viscosity change.

The engineering design of plant requires certain other factors which provide a suitable milk contact surface, cleaning without frequent dismantling, faster heat transfer and economy of steam/power used for operating the plant.

Following factors are important for evoparation process:

- Concentration: The initial and final concentration of solute in the solution should be considered. As the concentration increases, the boiling point rises.

- Foaming: Few products have tendency to foam, which reduce heat transfer and there is difficulty in controlling level of liquid which ultimately increases product (entrainment) losses.

- Heat sensitivity: Milk, like many other food products, is sensitive to high temperatures. If time of exposure is more, there will be severe damage to milk proteins.

- Scale formation / Fouling: It is a common phenomenon of deposition of solids on the heat exchanger surface. However, the scale forming tendency can be very much reduced by maintaining reasonably low temperature difference and relatively clean and smooth heat transfer surface. The flow velocity of product has also significant effect. If scale formation starts, rate of heat transfer decreases and cleaning becoming more difficult.

- Materials of construction: Stainless steel is the most common metal for evaporators in the dairy and food industry. Other metals may be used in chemical evaporators. The factors like strength, toughness, weld-ability, non-toxicity, surface finish, cost etc. are important in the selection of material of construction.

- Specific heat: It changes with concentration of solution. More heat is required to be supplied at high specific heat values.

- Gas liberation: Few products liberate gases when heated under boiling pressures.

- Toxicity: The gases liberated in few cases may be toxic and should be handle carefully.

- Viscosity: There is increase in viscosity of solution during evaporation which increases time of contact and hence chances of burning or damage the product.

- Capacity: It is expressed as the amount of water evaporated per hour. It depends on the surface area of heat transfer, temperature difference and the overall heat transfer co-efficient.

- Economy: It is based on the amount of water evaporated per kg of steam used. It increases with number of effects.

Different Types of Evaporators used in Dairy Industry

The major types of evaporators used in dairy industry are:

- Vertical tube circulation evaporator.

- Batch vacuum pan evaporator.

- Long tube vertical (rising and falling film type).

- Plate evaporators:

  ○ Film evaporators with mechanically moved parts (SSHE).

  ○ Expanding flow evaporator.

Different Types of Evaporators

Evaporators are of many different shapes, sizes and types of heating units. The major objective is to transfer heat from heat source to the product to evaporate water or other volatile liquids from the product. The general classification for evaporator bodies may be made based on:

- Source of heat,

- Position of tubes for heating,

- Method of circulation of product,

- Length of tube,

- Direction of flow of film of product,

- Number of passes,

- Shape of tube assembly for heat exchanger,

- Location of steam,

- Location of tubes.

The most important and widely used evaporator is the long tube vertical (calandria) type evaporator with climbing or falling film principle. The type is of the forced circulation type with steam condensing in the jacket surrounding a most of small diameter tubes. This type of evaporator has higher rate of heat transfer, less contact time with hot surface, flexibility of operation, economy of evaporation and easy-in-place cleaning. It can be operated in stages reusing vapours by Thermo-Vapour Recompression (TVR) and Mechanical Vapour Recompression (MVR), for steam economy.

Long Tube Vertical (Rising and Falling Film Type) Evaporator

In natural convection evaporators, the velocity of the fluid is usually less than one to 1.25 m/s. It is difficult to heat viscous materials with a natural circulation unit. Therefore the use of forced circulation to obtain a velocity of liquid up to 5 m/s, at the entrance of the tubes is desired for more rapid heat transfer. The liquid head above the heat exchanger is usually great enough to prevent boiling in the tubes. A centrifugal pump is normally used for circulation of milk products, but a positive pump is used for highly viscous fluids.

Tubes of 3 to 5 cm diameter and 300 to 500 cm long are used to move the liquid on the inside. These are placed in a steam chest. So that steam heats from the outside of the tube. The Long Tube Vertical (LTV) evaporator is used normally with the heating element separate from the liquid-vapour separator. The product enters at the bottom of the evaporator body and as it is heated by steam condensing on the opposite side of the tube, the product moves rapidly to the top of the tube and then into a separation chamber. The evaporator is thus a continuous one in operation. Within the tubes there are three distinct regions. At the bottom, under the static head of liquid, no boiling takes place, only simple heating occurs. In the center region the temperature rises sufficiently for boiling and vapour is produced, heat transfer rates are still low. In the upper region the volume of vapour increase and the remaining liquid is being wiped into a film on the tube surfaces resulting in good heat transfer conditions. The disadvantages of this type are the relatively large hold up of liquid in the lower regions of the tube giving long contact times (15-30 min.) Also evaporation ratio in a single pass is usually not sufficient to reach the required concentration, so that recycling is necessary, extending residence time. In the central portion of the tubes formation of scale, protein deposits and other fouling is often found to be most severe.

Vapour is removed by the separation chamber and the concentrated product removed or recirculated through the evaporation chamber again, depending on the concentration desired.

The falling film evaporator is used to reduce the amount of heat treatment and exposure of heat to the product. The tubes are from 4 to 5 cm diameter and up to 600 cm long in the falling film evaporator. The product is sprayed or other wise distributed over the inside of the tubes which are heated with steam. Unless the tubes are fairly heavily loaded there is a risk that some of the tubes may not get their fair share of feed and will overheat or over concentrate the liquid flowing down. The distributor is provided for uniform distribution of feed to each and every tubes of calandria to form thin film over the inner surface of tubes.

The smaller the tubes for a given output the easier it is to get even distribution, also small tubes result in a larger pressure drop across their length. The ideal plant might well have conical tubes which would maintain a good initial velocity, would prevent overloading with vapour at the bottom of the tubes and might make the distribution of the feed easier.

Moisture removed moves downward along with the concentrated product and finally separated in the vapour separator. The product may be recirculated for further concentration or removed from the system. The Reynolds number of the falling film should exceed 2000 for good heat transfer.

The great advantage of the falling film is the short time the product remains inside the tube. This gives better quality product with minimum changes or damage to the product. The other advantages are as stated below:

- The overall heat transfer coefficient is much larger than vacuum pan or other types of evaporator. The U-value of vacuum pan is 500-700 W/m² K, while for multi effect evaporators it is 1500-2200 W/m² K.

- More than one effect can be used in series with great saving in steam per kg of vapour.

- There is no static head and hence no change in the boiling point due to hydrostatic head.

- It can be used for concentrating most of the heat sensitive products including milk and fruit juices, due to lower temperature gradient.

- Evaporation is carried out at lower temperature due to higher vacuum and temperature difference required is relatively low.

Disadvantages are as under

- Chocking of tubes due to scale formation and difficulty in cleaning of tubes.

- Operation is highly sensitive to fluctuation in steam pressure to plant.

- Sudden failure of vacuum causes heavy entrainment losses and fouling of tubes.

- Great care is needed in keeping all joints leak proof to maintain desired vacuum.

Table: Difference between Rising and Falling Film Evaporators.

Rising Film Evaporators	Falling film Evaporators
More residence time.	Less residence time.
More temperature difference is required between heating medium and feed.	Less temperature difference is required between heating medium and feed.
Less overall heat transfer coefficient.	More overall heat transfer coefficient.
There is a static head and hence change in the boiling point due to hydrostatic head in the tube.	There is no static head and hence no change in the boiling point due to hydrostatic head in the tube.
Higher vacuum is not possible.	Higher vacuum is possible.
It is not used for heat sensitive products.	Used for heat sensitive products as gentle heating.
More fouling problem.	Less fouling problem.

Materials of Construction

Evaporator bodies and tubes are fabricated from the materials mostly of stainless steel (AISI-SS-316) is used when corrosive action is to be prevented.

Design Consideration

Evaporator drums invariably operate under vacuum. These are designed for an external pressure of 0.1 N/mm2 (100 kPa). The bottom head may be conical in many cases and may be designed for similar pressure rating. The top head may be flanged for flared and dished shape or conical. The calandria which has the tubular heating surface is designed as a shell and tube heat exchanger. Since steam under pressure in usually accepted as the heating medium, the design is based on the pressure of steam. The entire evaporator body must be rigid. The conical head, the calandria and the vapour drum are connected by flanged joints or directly welded. The vapour drum may be made up to separate cylindrical pieces and joined by flanges. Large openings like manholes, sight glasses must be reinforced with compensating rings. Supports may be placed below the brackets welded to the vapour drum or to the calandria. External calandria is also designed as a shell and tube heat exchanger.

Plate Evaporator

The plate evaporator is characterized by a large heat exchanger surface occupying a relatively small space which need not be very high. Like the plate heat exchanger, it is constructed from profile plates, with the condensing steam used as heating medium and the evaporating product passing between alternate pairs.

The advantages of plate evaporator are its, flexibility, low head space, sanitary construction and shorter residence time which makes evaporation of heat sensitive products possible. It also offers possibility of multiple effects. However, rubber gaskets for sealing are costly; Liquid having suspended matter cannot be easily processed. For even distribution and to ensure good wetting of the surface, orifice pieces are to be inserted at header ports. Sometimes recirculation is necessary to ensure proper wetting.

Film Evaporators with Mechanically Moved Parts (SSHE)

When highly viscous products (viscosity more than 1 Pa s) or fluids containing suspended matter are to be evaporated, it may happen that the forces which normally move the liquid along with gravity and propelling power of the vapour, are not sufficient to move the product satisfactorily. This intensifies the problem of maintaining high rates of heat transfer and proper distribution.

A shaft fitted with wiper blades, scrapers, vanes or other device rotates within a vertical tube of relatively large diameter. This tube is surrounded by a heating jacket. The rotor may have a fixed clearance of 0.2 – 2.0 mm or fixed blades with adjustable clearance, or blades which actually wipe the heat exchanger surface. The purpose of the blades etc. is to produce thorough mixing of the film, to distribute it evenly and to transport the product through the evaporator. The film thickness differs from one liquid to another depending on its physical properties.

The advantages of this evaporator are:

- It can handle highly viscous, pulpy and foaming materials.

- Evaporation rates are high.

- Fouling problem non-existent.

The disadvantages are:

- Requires precise alignment because of small blade clearance.

- Difficult to clean.

- High capital and operating cost.

- High headspace required for demounting rotor for inspection and cleaning.

Expanding Flow Evaporator

It is compact and its heating element and expansion vessel are a single unit. In put milk acts as coolant in condenser. Steam condensate is used in milk pre heater. CIP is possible. Flexible in its capacity. One can get concentration in one pass. It has shorter residence time of < 1 min. Hence it is giving the advantage of gentle heating. Also because of low holding the plant has the characteristic of quick start up. It is made up of number of inverted, S.S. cones. Gaskets maintain narrow passages between cones. The alternative passages for feed and steam is provided.

Entrainment Separators

The entrainment separators are basically depend on principles of impingment theory, where liquid droplets get returned due to spiral tubes and baffles installed in the path of the vapour. The other principle is change in direction as well as velocity. For industrial applications centrifugal type of entrainment separators are in use.Entrainment results whenever a vapour is generated from a liquid. So the vapour carries these liquid droplets. Separators provide a separation of liquid from vapour.

Vapour Release Chamber

A large chamber is used to reduce the velocity of vapour stream. This enables the droplets to settle out by gravity. The vapour release drum may either be placed just above the bundle shell or it is a separate unit placed adjoining to tube bundle shell, being connected to it by a large pipe. It may not be economically practical to make the vapour head large enough to accomplish the entire decontamination of the vapour. Further, increasing vapour space decreases entrainment of larger drops, but has not effect on small drops.

The vapour disengagement rate from a boiling liquid surface should not normally be more than 30 cm per second for normal solutions at atmospheric pressure and may be about 3 cm per second with crude solutions. Even allowing for sufficient vapour disengagement space it is common practice to provide spray traps. These traps are merely a series of baffles giving rapid changes of direction to the vapour stream.

Wire Pad

Pads of finely woven wire set in the vapour release chamber at right angles to the vapour flow are used for entrainment. As the vapour and its entrained liquid pass through the pad, the liquid particles agglomerate, eventually falling back into the vapour release chamber. A highly purified dry vapour leaves the top of the pad. Application of such pads may be difficult for vapours with suspended solids, fibers or scale forming materials, which will block the wire mesh. In such cases washing facilities at proper intervals may be provided. Wire pads are not generally used in the food industry for the unhygienic condition it creates.

Vapour Release Drum Size

The size of drum provided above the tube bundle in most of evaporators, is decided by three important considerations. They are: (a) the foaming of the liquid in the evaporator, (b) the vapour velocity, and (c) entrainment separation. Foaming takes place above the liquid level and occupies a certain space of the vapour drum. The vapour velocity sets the minimum drum diameter.

A thumb rule commonly employed in evaporator design of this kind is that the height of the vapour space above the calandria should not be less than one vessel diameter and the bottom space below calandria should be one-fourth of vessel diameter. In cases where the entrainment separator forms an independent unit, the main drum can have a shorter disengaging height.

Centrifugal Separator

This is a separate drum in which the vapours are admitted tangentially and are made to flow in a helical path by use of baffles. The vapours leave either from the top of the drum or through a central pipe.

High heat transfer coefficients are obtained and viscous materials are handled at relatively high temperature but for shorter contact times. The plate arrangement may be such that it offers a combination of rising and falling film principle or falling film principle alone. By varying the plate gap, width of the plates and the relative dimensions of the various channels, the vapour velocity is controlled for efficient heat transfer. High heat transfer, larger cross-sectional areas are provided for the inlet of the steam used for heating than for the discharge of condensate. Similarly, the cross-sectional areas for discharge of vapour and of concentrate are also enlarged.

PACKAGING MACHINES FOR MILK

The Form Fill Seal(FFS) machines are a genre of filling equipment that can fill in a flexible packing material. The product should be free flowing type, either liquid or even granular. The equipment may be controlled electro-pneumatically or mechanically.

The process involves certain steps, which will take place cyclically in auto operation.

There is option for variation in size and quantity of the product packed. To a large extent, the market milk is now being sold by packing in these machines.

The following are the operations that go cyclically:

- Forming of tube of packing material from film in rolled state.

- Simultaneous operation of filling and sealing.

- Movement of film to form next package.

- Simultaneous separation of filled and sealed packet while filling of next packet.

Sequence of Operation

The film roll is loaded at the backside of the filling machine on a sliding platform. The film edge is passed over end of role contact lever, dancer roller, UV tube and brought to the front side over to forming plates. The forming plates rolled the flat film in to tube a certain band of overlap. Within the tube is the fluid filling pipe enveloped. The tube then passes over to vertical seal jaws that are engaged and disengaged with the help of an air operated piston, or in machines by mechanical means. In between the jaws, the overlapped part of the film tube passes. The set of jaws have one stationary and one moving jaw. The moving jaw has a nichrome rod, supplied with variable voltage such that the heat is generated when the current passes intermittently. During the period when the current does not pass, when the jaws are disengaged, the cooling water being circulated in the moving jaw, cools it and prevents continuous over heating of the sealing rod. The film is supported by Teflon cloth and rubber cushion, as well as protected by Teflon cloth from sealing rod. This arrangement prevents electricity passing on to the film and other parts, while allowing only the heat to pass on to the film and partly melting and fusing the vertical joint.

Lower down the film tube, there are a pair of nip rollers giving a holding and pulling down action, when the jaws are disengaged, making the film to move to seal the next portion of vertical overlap.

Further lower down the film tube is engaged by horizontal jaws, at a sufficiently below the lower edge of fluid filling pipe. This arrangement allows the formation of lower seal of the packet, while the fluid is being filled to a known quantity. The quantity of flow is controlled by a valve operated by a rod which is lifted by a solenoid coil position at the top of machine, just at the feeding line from the over head tank carrying the fluid to be filled. While filling is taking place, a pair of flat blades operated by spring keeps the film perfectly flat at horizontal edge so that there is no folds and horizontal seal is perfect.

When the filling of fluid and the horizontal sealing is complete, the horizontal jaws (as well as vertical jaws) get disengaged, and the nip rollers start rolling to bring the next length of film tube to be filled for next packet. While the second packet is being filled, the first packet already filled will be getting the horizontal seal of top portion of the filled packet. When the next time the jaws open, the first packet drops down by its own weight and weakened connection to the rest of the tube.

The above cycle of operation is repeated when the controls are in automatic operation, while single action takes place when in manual operation during initial adjustment of time and temperature combination for obtaining proper seal.

Controls

The Form Fill Seal Machine has various controls for the following operations:

- Adjusting the temperature of sealing rod by controlling the electric supply, to match to the thickness of the film to be sealed.

- Adjusting the timing for the jaws to be engaged and simultaneously filling operations to take place, with a known quantity of fluid.

- Adjusting the quantity of fluid to be filled when jaws are engaged.

- Adjusting the timing for the jaws to be engaged and allow time for movement of film to the required length of package.

The other useful instrumentation are the end of film indicator (gives audio signal), so that the new film roll can be changed easily, fine adjustments for the quantity of fluid filled in few grams range, fine adjustment of timings, etc.

The equipment will require water for cooling the jaws at a fixed flow rate and of low temperature. Also, compressed air is required which is at required pressure and free of condensed moisture.

UHT Milk Packing Machine

The UHT milk packing machine is different than the usual FFS machine in that the packaging material is multi layered, and the filling is done in an absolutely asceptic condition. The filling room is kept in a positive pressure, and the air inlet is through HEPA filters. Starting out from a reel of packaging material, the Tetra Brick Asceptic (TBA) filling machine produces filled packages. The packaging material is first sterilized and then formed into a tube. The tube is filled with product and then shaped and cut into individual packages.

The Package

There is a range of Asceptic packages , all deriving their origin from the same forming technique.

Package Terminology

Creases are the folding instructions on the packaging material, to ensure the Creases package's final shape. The creases are pressed into the material by the creasing tools in the converting process.

The longitudinal seal (LS) is accomplished when forming the packaging material seals into a tube. It seals the package along the side. A strip of laminated plastic, the LS-strip, covers the seal on the inside. The area of the overlap joint is called the longitudinal overlap.

The transversal seal (TS) is made when the tube is filled with product. It seals the package at top and bottom. The sealing takes place below the product level in the tube.

The fins are the areas, at top and bottom of the package, where it is sealed and Fins cut.

The flaps would be the corners of the package, if you flattened it out. When Flaps shaping the package, the flaps are folded down and in, and then sealed to the package body.

Machine Introduction

Tetra Brik filling machines are built from so called modules or main groups with similar functions in the various machines. The machines may also have different additional equipment and accessories.

- ASU Automatic splicing unit(ASU): The automatic splicing unit splices reels of packaging material. This means that production can continue uninterrupted when one reel of packaging material comes to an end.

- During splicing though, the packaging material has to remain still in the splicing head. The magazine provides the necessary supply of material so that the machine does not have to stop.

Pull Tab Unit

- The Pull Tab unit is additional equipment, providing the packaging material with a Pull Tab opening before it enters the peroxide bath. The Pull Tab opening is created by punching a hole in the packaging material. The hole is sealed with plastic on the inside and aluminium on the outside.

- Strip applicator: The strip applicator applies a plastic strip, the LS-strip, along one edge of the packaging material. The strip is applied on the inside of the packaging material and is intended to prevent product from being soaked into the raw paper edge of the longitudinal seal. The strip will also support the seal. Only half of the LS-strip is sealed to this edge of the packaging material. The other half will be sealed to the other edge later, when the packaging material is formed into a tube.

- Peroxide bath: The packaging material will be sterilized in the peroxide bath. In machines with deep baths, as shown in the example, the packaging material will be immersed into warm peroxide and both sides will be sterilized. In machines with shallow baths the inside of the packaging material will merely be covered with cold peroxide and the sterilization will be finished in the tube heater.

- Aseptic chamber: The packaging material will be dried with heated air. In machines with deep baths, as shown in the example, an aseptic environment around the sterilized packaging material is maintained with an overpressure of heat-sterilized air. This takes place in the aseptic chamber. Machines with shallow baths, which have no aseptic chamber, heat-sterilized air will be blown into the tight tube. This way a sterile area is maintained where the tube is to be filled with product. The packaging material will be formed into a tube and sealed longitudinally. Finally, the tube will be filled with product.

- Jaw system: In the jaw system the tube is sealed transversally and cut into separate packages. The sealing is made by induction heating, using the aluminum in the packaging material to melt the plastic. It is important that the package design, with the creases, appear in accordance with the jaws. This is controlled and corrected by the jaw system.

- Final folder: In the final folder the separate package gets its final shape. The fins are folded and the flaps are folded and sealed. Hot air is used to seal the flaps. The plastic outer coating on the package material is heated and the flaps are pressed against the sides and the bottom of the package. When the plastic gets cool the flap is sealed.

- Operator panel: The operator panel allows the operator to communicate with the machine. It is used to start and stop or make the machine take any other action.

- Electrical cabinet: In the electrical cabinet a great part of the electrical components are included, such as:

  ○ Temperature regulators,

  ○ Control system,

  ○ Contactors,

  ○ Induction Heating unit, etc.

- Service unit: The service unit includes parts and supply systems needed for the machine function, for example:

  ○ Water and air system,

  ○ Lubrication and hydraulic oil system,

  ○ Pneumatic and peroxide systems.

- Drive system: The drive system includes motor, gear and cam. These parts run the jaw system and also the final folder on certain machines.

The wide varieties of milk products packaging makes the choice of packing machines. The requirement of each product is different,as well as the size and shape. Some of them are to be stored at low temperatures, while others are kept in ambient temperatures. The packing material also different, and the information to be given on packages may be a legal requirement in most of the cases. Some products are further processed while in the packaging itself, like sterilization, curd setting etc. The packaging machine working principle will also depend on the physical nature of the product, like liquid, solid, semi solid.

Cheese Packing

Specific Applications

When packaging food products those are sensitive to spoilage or loss of freshness, it is important to seal the product in containers that allow a tight seal. It is typical to evacuate the chamber to a target vacuum level, thus removing air and oxygen. Higher vacuum levels remove more air, but require larger vacuum systems, higher attainable vacuum levels, and more time for the evacuation process. The target vacuum level is a function of the product being packaged and is determined by the packager and/or government regulations. Speed is an important factor in many packaging operations, and machine manufacturer are often rated on how many packs per hour their machine can perform, while meeting a set of specifications. Higher speed usually means higher vacuum system capacities, and/or a secondary vacuum system to "pre-evacuate" the vacuum packaging chamber. Sometimes, purge gases are introduced during the packaging operation to help displace air and oxygen, retard spoilage, and further enhance shelf life and appearance. Appearance is vitally important in the packaging industry.

Machine Types

Chamber Packaging Machines

Chamber machines are typically used to seal meats and cheeses in film pouches. In appearance, they have a base and a swinging top cover (sort of like a Xerox machine). The product is placed into the machine, and the swing lid is closed.Where the lid and top of the machine meet, a seal is formed. At this point vacuum evacuation begins and the rest of the sealing process occurs under vacuum. When the lid is lifted, the product is fully packaged and ready to remove.

Small Chamber Machine (Desk Model)

Chamber up to 50 liters, rotary vane pump 3 to 25 m³/hr.

Double Chamber (Mobile)

Two chambers, each up to 150 liters, rotary vane pump 40 to 300 m³/hr.

Big Chamber Machine (Non Nobile)

Chamber up to 400 liters, pump combination rotary vane pump 340 to 1000 m3/hr.Booster pumps 500 to 2000 m³/hr.

Rotating Chamber Machine (Non Mobile)

Multiple rotating chambers (step evacuation) rotary vane pump160 to 630 m³/hr.

Multiple rotating chambers (step evacuation) rotary vane pump 160 to 630 m³/hr and booster pump 500 to 2000 m³/hr.

Tray Sealing Machine

With Tray type machines, the user purchases a tray (Styrofoam, plastic, etc.), loads the product onto the trays and places them into the machine. At this point the chamber is evacuated to achieve a "skin pack". These machines are typically used by small packagers.

Inert Gas-flushed Packaging (Snorkel Machines)

MAP(Modified Atmosphere Packaging) involves packaging the food product in an gas atmosphere other than standard room air to achieve specific goals not otherwise possible. Common atmospheres include:

- Nitrogen,

- 80% Nitrogen- 20% CO_2 mixture,

- Saturated oxygen greater then 80%.

The machine pulls a vacuum and reduces oxygen content down to 0.3 to 1% to remove Anabolic Bacteria. When this occurs, meat becomes purple. Back flushing with the appropriate gas maintains product color for an attractive and appetizing appearance. All machines that use a chamber can be fitted with gas flush option, and approximately 25% of machines on the market do so.

INDEX

www.ingramcontent.com/pod-product-compliance
Lightning Source LLC
Chambersburg PA
CBHW082058190326
41458CB00010B/3526